OVERPOWERED

OVERPOWERED

What Science Tells Us about the Dangers of Cell Phones and Other WiFi-Age Devices

Martin Blank, PhD

Seven Stories Press
NEW YORK

A Seven Stories Press First Edition

Seven Stories Press
140 Watts Street
New York, NY 10013
www.sevenstories.com

College professors and high school and middle school teachers may order free examination copies of Seven Stories Press titles. To order, visit www.sevenstories.com/textbook or send a fax on school letterhead to (212) 226-1411.

Library of Congress Cataloging-in-Publication Data
Blank, Martin, 1933-
Overpowered : the dangers of electromagnetic radiation and what you can do about it / Dr. Martin Blank, PhD. -- A Seven Stories Press first edition.
pages cm
ISBN 978-1-60980-509-8 (hardback)
1. Electromagnetic waves--Health aspects--Standards. 2. Electromagnetic fields--Safety measures. 3. Cell phones--Health aspects. 4. Nonionizing radiation--Safety measures--Standards. I. Title.
RA569.3.B58 2014
539.2--dc23

2013024601

Printed in the United States

9 8 7 6 5 4 3 2 1

CONTENTS

PREFACE

At various gatherings, when I am inevitably asked what I do for a living, an interesting sequence often ensues. "I am working on the biological effects of cell phones, WiFi, and related devices," I reply. That is followed by a slightly anxious, "So, are they dangerous?" After I say that there is a considerable body of evidence showing significant risk, the conversation usually ends—often with a statement such as, "Well there's no way I am giving up my cell phone."

These experiences have shown me that in writing this book, I face a bit of an uphill battle. While people are on edge about the health issues, they are also on edge about the possibility of having to give up the amazing technologies that have become so much a part of their lives. Allow me to put you at ease right from the start: There is no need for us to abandon the devices of the electronic age.

But there is a vast continuum of ways to address the problems—from abandonment to the unrestricted use that currently exists. One has only to look at the story of aerosols to see how safer alternatives come to light once a problem is identified. When aerosols first appeared, they were seen as miracle substances, as is often the case with many technological advances. Then in the 1970s, studies showed that the propellants in aerosols were depleting the ozone layer that is critical to life on earth. National and international organizations were called to action, and there was an expanding and effective ban on the use of the most dangerous propellants. As a result, today we still have aerosols, but the emission of ozone-depleting substances is on the decline.

Whenever arguments are raised about implementing limitations, the typical corporate-sponsored line is that "there is no solid evidence of danger." I have written this book to show you that this is not the case.

There is a large body of solid science showing that the electromagnetic radiation (EMR) that is a by-product of our high-tech world has many and varied effects on our biology. It is time we replaced the common refrain of "no solid evidence of danger" with "it's time we acknowledge the dangers and do something about it."

The steps necessary for change are many. Two of the most important are:

1. Establish the criteria that will make the technology safer for users and for all in the surrounding environment. Fortunately, those criteria are not difficult to put in place, and there are multiple ways in which the amount of exposure can be reduced. What is required is the recognition that the change is needed and doable.
2. Create an informed citizenry. When people are informed, they are empowered. We have seen over and over again what the population can achieve when it is mobilized to act. Corporate and government policy can change dramatically.

These goals are central to the ideas in this book. By knowing the facts, you can make informed decisions about how you use technology, and you can also become part of the process required to reduce the potential harm posed by EMR.

Chapter 1

AN UNLIKELY ACTIVIST

You may not realize it, but you are participating in an unauthorized experiment—"the largest biological experiment ever," in the words of Swedish neuro-oncologist Leif Salford. For the first time, many of us are holding high-powered microwave transmitters—in the form of cell phones—directly against our heads on a daily basis.

Cell phones generate electromagnetic fields (EMF), and emit electromagnetic radiation (EMR). They share this feature with all modern electronics that run on alternating current (AC) power (from the power grid and the outlets in your walls) or that utilize wireless communication. Different devices radiate different levels of EMF, with different characteristics.

What health effects do these exposures have?

Therein lies the experiment.

The many potential negative health effects from EMF exposure (including many cancers and Alzheimer's disease) can take decades to develop. So we won't know the results of this experiment for many years—possibly decades. But by then, it may be too late for billions of people.

Today, while we wait for the results, a debate rages about the potential dangers of EMF. The science of EMF, discussed in the next chapter, is not easily taught, and as a result, the debate over the health effects of EMF exposure can get quite complicated. To put it simply, the debate has two sides. On the one hand, there are those who urge the adoption of a precautionary approach to the public risk as we continue to investigate the health effects of EMF exposure. This group includes many scientists, myself included, who see many danger signs that call out strongly for precaution. On the other side are those who feel that we should wait for definitive proof of harm before taking any action. The most vocal of this group include representatives of industries who undoubtedly perceive threats to their profits and would prefer that we continue buying and using more and more connected electronic devices.

This industry effort has been phenomenally successful, with widespread adoption of many EMF-generating technologies throughout the world. But EMF has many other sources as well. Most notably, the entire power grid is an EMF-generation network that reaches almost every individual in America and 75% of the global population. Today, early in the 21st century, we find ourselves fully immersed in a soup of electromagnetic radiation on a nearly continuous basis.

WHAT WE KNOW

The science to date about the bioeffects (biological and health outcomes) resulting from exposure to EM radiation is still in its early stages. We cannot yet predict that a specific type of EMF exposure (such as 20 minutes of cell phone use each day for 10 years) will lead to a specific health outcome (such as cancer). Nor are scientists able to define what constitutes a "safe" level of EMF exposure.

However, while science has not yet answered all of our questions, it has determined one fact very clearly—*all electromagnetic radiation impacts living beings*. As I will discuss throughout this book, science demonstrates a wide range of bioeffects linked to EMF exposure. For instance, numerous studies have found that EMF damages and causes mutations in DNA—the genetic material that defines us as individuals and collectively as a species. Mutations in DNA are believed to be the initiating steps in the development of cancers, and it is the association of cancers with exposure to EMF that has led to calls for revising safety standards. This type of DNA damage is seen at levels of EMF exposure equivalent to those resulting from typical cell phone use.

The damage to DNA caused by EMF exposure is believed to be one of the mechanisms by which EMF exposure leads to negative health effects. Multiple separate studies indicate significantly increased risk (up to two and three times normal risk) of developing certain types of brain tumors following EMF exposure from cell phones over a period of many years. One review that averaged the data across 16 studies found that the risk of developing a tumor on the same side of the head as the cell phone is used is elevated 240% for those who regularly use cell phones for 10 years or more. An Israeli study found that people who use cell phones at least 22 hours a month are 50% more likely to develop cancers of the salivary gland (and there has been a four-fold increase in the incidence of these types of tumors in Israel between 1970 and 2006).[1] And individuals who lived within 400 meters of a cell phone transmission tower for 10 years or more were found to have a rate of cancer three times higher than those living at a greater distance.[2] Indeed, the World Health Organization (WHO) designated EMF—including power frequencies and radio frequencies—as a possible cause of cancer.

While cancer is one of the primary classes of negative health effects studied by researchers, EMF exposure has been shown to increase risk for many other types of negative health outcomes. In fact, levels of EMF thousands of times lower than current safety standards have been shown to significantly increase risk for neurodegenerative diseases (such as Alzheimer's and Lou Gehrig's disease) and male infertility associated with damaged sperm cells. In one study, those who lived within 50 meters of a high voltage power line were significantly more likely to develop Alzheimer's disease when compared to those living 600 meters or more away. The increased risk was 24% after one year, 50% after 5 years, and 100% after 10 years.[3] Other research demonstrates that using a cell phone between two and four hours a day leads to 40% lower sperm counts than found in men who do not use cell phones, and the surviving sperm cells demonstrate lower levels of motility and viability.

EMF exposure (as with many environmental pollutants) not only affects people, but all of nature. In fact, negative effects have been demonstrated across a wide variety of plant and animal life. EMF, even at very low levels, can interrupt the ability of birds and bees to navigate. Numerous studies link this effect with the phenomena of avian tower fatalities (in which birds die from collisions with power line and communications towers). These same navigational effects have been linked to colony collapse disorder (CCD), which is devastating the global population of honey bees (in one study, placement of a single active cell phone in front of a hive led to the rapid and complete demise of the entire colony[4]). And a mystery illness affecting trees around Europe has been linked to WiFi radiation in the environment.

As I explain in the coming chapters, there is a lot of science—high-quality, peer-reviewed science—demonstrating these and other very troubling outcomes from exposure to electromagnetic radiation. These effects are seen at levels of EMF that, according to regulatory agencies like the Federal Communications Commission (FCC), which regulates cell phone EMF emissions in the United States, are completely safe.

AN UNLIKELY ACTIVIST

I have worked at Columbia University since the 1960s, but I was not always focused on electromagnetic fields. My PhDs in physical chemistry from Columbia University and colloid science from the University of Cambridge provided me with a strong, interdisciplinary academic background in biology, chemistry, and physics. Much of my early career was spent investigating the properties of surfaces and very thin films, such as those found in a soap bubble, which then led me to explore the biological membranes that encase living cells.

I studied the biochemistry of infant respiratory distress syndrome (IRDS), which causes the lungs of newborns to collapse (also called hyaline membrane disease). Through this research, I found that the substance on the surface of healthy lungs could form a network that prevented collapse in healthy babies (the absence of which causes the problem for IRDS sufferers).

A food company subsequently hired me to study how the same surface support mechanism could be used to prevent the collapse of the air bubbles added to their ice cream. As ice cream is sold by volume and not by weight, this enabled the company to reduce the actual amount of ice cream sold in each package. (My children gave me a lot of grief about that job, but they enjoyed the ice cream samples I brought home.)

I also performed research exploring how electrical forces interact with the proteins and other components found in nerve and muscle membranes. In 1987, I was studying the effects of electric fields on membranes when I read a paper by Dr. Reba Goodman demonstrating some unusual effects of EMF on living cells. She had found that even relatively weak power fields from common sources (such as those found near power lines and electrical appliances) could alter the ability of living cells to make proteins. I had long understood the importance of electrical forces on the function of cells, but this paper indicated that magnetic forces (which are, as I will explain in the next chapter, a key aspect of electromagnetic fields) also had significant impact on living cells.

Like most of my colleagues, I did not think this was possible. By way of background, there are some types of EMF that everyone had long acknowledged are harmful to humans. For example, X-rays and ultraviolet radiation are both recognized carcinogens. But these are *ionizing* forms of radiation. Dr. Goodman, however, had shown that even *non-ionizing* radiation, which has much less energy than X-rays, was affecting a very basic property of cells—the ability to stimulate protein synthesis.

Because non-ionizing forms of EMF have so much less energy than ionizing radiation, it had long been believed that non-ionizing electromagnetic fields were harmless to humans and other biological systems. And while it was acknowledged that a high enough exposure to non-ionizing EMF could cause a rise in body temperature—and that this temperature increase could cause cell damage and lead to health problems—it was thought that low levels of non-ionizing EMF that did not cause this rise in temperature were benign.

In over 20 years of experience at some of the world's top academic institutions, this is what I'd been taught and this is what I'd been teaching. In fact, my department at Columbia University (like every other comparable department at other universities around the world) taught an entire course in human physiology without even mentioning magnetic fields, except when they were used diagnostically to detect the effects of the electric currents in the heart or brain. Sure magnets and magnetic fields can affect pieces of metal and other magnets, but magnetic fields were assumed to be *inert*, or essentially powerless, when it came to human physiology.

As you can imagine, I found the research in Dr. Goodman's paper intriguing. When it turned out that she was a colleague of mine at Columbia, with an office just around the block, I decided to follow up with her, face-to-face. It didn't take me long to realize that her data and arguments were very convincing. So convincing, in fact, that I not only changed my opinion on the potential health effects of magnetism, but I also began a long collaboration with her that has been highly productive and personally rewarding.

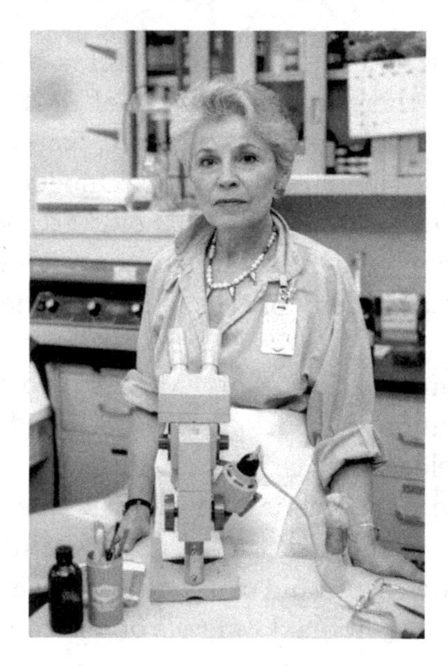

Reba Goodman, PhD Professor Emeritus, Clinical Pathology
Courtesy of Columbia University Medical Center

During our years of research collaboration, Dr. Goodman and I published many of our results in respected scientific journals. Our research was focused on the cellular level—how EMF permeate the surfaces of cells and affect cells and DNA—and we demonstrated several observable, repeatable health effects from EMF on living cells. As with all findings published in such journals, our data and conclusions were peer reviewed. In other words, our findings were reviewed prior to publication to ensure that our techniques and conclusions, which were based on our measurements, were appropriate. Our results were subsequently confirmed by other scientists, working in other laboratories around the world, independent from our own.

A CHANGE IN TONE

Over the roughly 25 years Dr. Goodman and I have been studying the EMF issue, our work has been referenced by numerous scientists, activists, and experts in support of public health initiatives including the *BioInitiative Report* (discussed in chapter 11), which was cited by the European Parliament when it called for stronger EMF regulations. Of course, our work was criticized in some circles, as well. This was to be expected, and we welcomed it—discussion and criticism is how science advances. But in the late 1990s, the criticism assumed a different character, both angrier and more derisive than past critiques.

On one occasion, I presented our findings at a US Department of Energy annual review of research on EMF. As soon as I finished my talk, a well-known Ivy League professor said (without any substantiation) that the data I presented were "impossible." He was followed by another respected academic, who stated (again without any substantiation) that I had most likely made some "dreadful error." Not only were these men wrong, but they delivered their comments with an intense and obvious hostility.

I later discovered that both men were paid consultants of the power industry—one of the largest generators of EMF. To me, this explained the source of their strong and unsubstantiated assertions about our research. I was witnessing firsthand the impact of private, profit-driven industrial efforts to confuse and obfuscate the science of EMF bioeffects.

NOT THE FIRST TIME

I knew that this was not the first time industry opposed scientific research that threatened their business models. I'd seen it before many times with tobacco, asbestos, pesticides, hydraulic fracturing (or "fracking"), and other industries that paid scientists to generate "science" that would support their claims of product safety.

That, of course, is not the course of sound science. Science involves generating and testing hypotheses. One draws conclusions from the available, observable evidence that results from rigorous and reproducible experimentation. Science is not sculpting evidence to support your existing beliefs. That's propaganda. As Dr. Henry Lai (who, along with Dr. Narendra Singh, performed the groundbreaking research demonstrating DNA damage from EMF exposure discussed at greater length in chapter 4 and elsewhere in this book) explains, "a lot of the studies that are done right now are done purely as PR tools for the industry."[5]

AN IRREVERSIBLE TREND

Of course EMF exposure—including radiation from smart phones, the power lines that you use to recharge them, and the other wide variety of EMF-generating technologies—is not equivalent to cigarette smoking. Exposure to carcinogens and other harmful forces from tobacco results from the purely voluntary, recreational activity of smoking. If tobacco disappeared from the world tomorrow, a lot of people would be very annoyed, tobacco farmers would have to plant other crops, and a few firms might go out of business, but there would be no additional impact.

In stark contrast, modern technology (the source of the human-made electromagnetic fields discussed in this book) has fueled a remarkable degree of innovation, productivity, and improvement in the quality of life. If tomorrow the power grid went down, all cell phone networks would cease operation, millions of computers around the world wouldn't turn on, and the night would be illuminated only by candlelight and the moon—we'd have a lot less EMF exposure, but at the cost of the complete collapse of modern society.

EMF isn't just a by-product of modern society. EMF, and our ability to harness it for technological purposes, is the *cornerstone* of modern society. Sanitation, food production and storage, health care—these are just some of the essential social systems that rely on power and wireless communication. We have evolved a society that is fundamentally reliant upon a set of technologies that generate forms and levels of electromagnetic radiation not seen on this planet prior to the 19th century.

As a result of the central role these devices play in modern life, individuals are understandably predisposed to resist information that may challenge the safety of activities that result in EMF exposures. People simply cannot bear the thought of restricting their time with—much less giving up—these beloved gadgets. This gives industry a huge advantage because there is a large segment of the public that would rather not know.

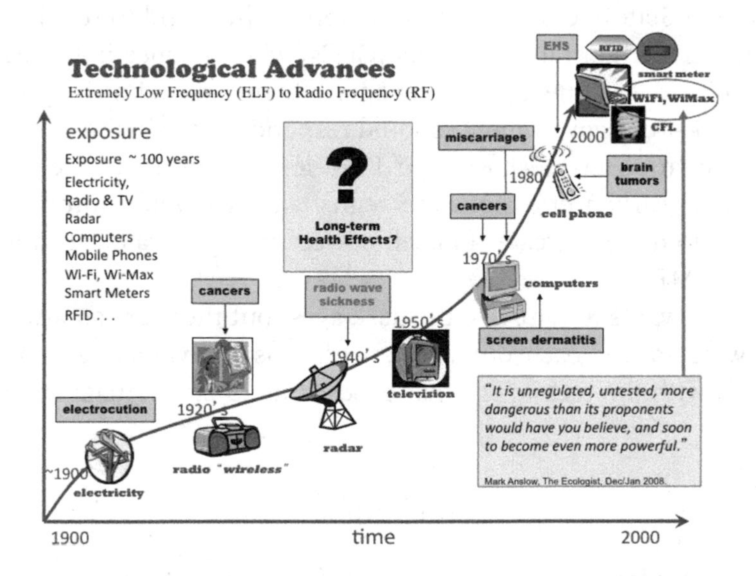

From 'Public Health SOS' by Camilla Rees and Magda Havas, with permission.

PRECAUTION

My message in this book is *not* to abandon gadgets—like most people, I too love and utilize EMF-generating gadgets. Instead, I want you to realize that EMF poses a real risk to living creatures and that industrial and product safety standards must and can be reconsidered. The solutions I suggest in this book are not prohibitive. I recommend that as individuals we adopt the notion of "prudent avoidance," minimizing our personal EMF exposure and maximizing the distance between us and EMF sources when those devices are in use. Just as you use a car with seat belts and air bags to increase the safety of the inherently dangerous activity of driving your car at a relatively high speed, you should consider similar risk-mitigating techniques for your personal EMF exposure.

On a broader social level, adoption of the Precautionary Principle in establishing new, biologically based safety standards for EMF exposure for the general public would be, I believe, the best approach. Just as the United States became the first nation in the world to regulate the production of chlorofluorocarbons (CFCs) when science indicated the threat to earth's ozone layer—long before there was definitive proof of such a link—our governments should respond to the significant public health threat of EMF exposure. If EMF levels were regulated just as automobile carbon emissions are regulated, this would force manufacturers to design, create, and sell devices that generate much lower levels of EMF.

No one wants to return to the dark ages, but there are smarter and safer ways to approach our relationship—as individuals and across society—with the technology that exposes us to electromagnetic radiation.

THIS BOOK

I have always looked to science as a reliable means of understanding a problem and as a source of information about possible solutions. My training, career, and belief in science have shown me that ultimately, knowledge is power. In the field of EMF, that knowledge is steadily growing.

The EMF issue spans physics, biology, and chemistry (as well as electrical engineering). My interdisciplinary background—which includes almost all of these fields—provides me with a valuable perspective on the issue of EMF exposure and its effect on living beings. This is what I've set out to relate in this book.

In the coming chapters, I will attempt to summarize and simplify the significant amount of information that I've learned about the health effects of EMF over the course of my career. (Those of you interested in more detailed science on these issues may review the materials referenced in the endnotes.) My goal is to demonstrate that all EMF—even at very low levels initially considered harmless—affects living beings. And the types of EMF exposures that result from increasingly common activities such as making a cell phone call, or using WiFi to access the Internet are linked with some very serious public health risks.

With this knowledge, I hope that you will do more to protect yourself and your family, work to reduce unnecessary dangers in your community, and ultimately be an informed consumer of the technology that surrounds you.

Chapter 2

ELECTROMAGNETIC FIELDS

This chapter explains the physical properties of electromagnetic radiation, the different types that exist, and the units of measurement—such as watts and volts—that are used throughout the remainder of this book. If you would rather skip this chapter, doing so will not impair your ability to read and appreciate any of the book's remaining content.

Back in 1998, four men were just finishing up the 16th hole at a golf course in Colorado, when, without warning, an electrical storm erupted. Seemingly out of nowhere, lightning struck a tree under which the men were standing. One of the golfers sustained serious burns. Two were knocked unconscious. The fourth suffered no external burns nor showed any evidence of having been hit by lightning. Yet, puzzling his doctors, he suffered cardiac arrest and died about three weeks later.

How could something like this occur? How could a man be affected by lightning when the lightning did not actually make contact with his body? As researchers concluded in the June 13, 1998, issue of the *Lancet*, the answer is EMF, or electromagnetic fields.

Electromagnetic fields are invisible forces that surround us—increasingly so in our modern, electrically powered world. The underlying science of electromagnetic radiation can be complex. In this chapter, I'll break down the most important concepts in EMF to enhance your understanding of the issues discussed in this book. Electromagnetic fields (as the name implies) emerge from a combination of two commonly experienced and well-understood forces in nature: electricity and magnetism.

ELECTRICITY

All matter is composed of atoms (such as carbon and iron), and all atoms are made up of the same fundamental particles that are negative (electrons), positive (protons), or uncharged (neutrons). The particles in an atom are in the form of a solar system with a nucleus containing the much heavier protons and neutrons in the center, and the much lighter electrons orbiting around the nucleus. A model of an atom of lithium is shown below with three electrons orbiting around the nucleus containing three protons. Different substances are characterized by the number of protons in the nucleus (e.g., carbon has 6 and iron has 26). Since atoms have the same number of electrons as protons, they are uncharged. However, electrons are relatively light and atoms can easily gain or lose electrons to become charged.

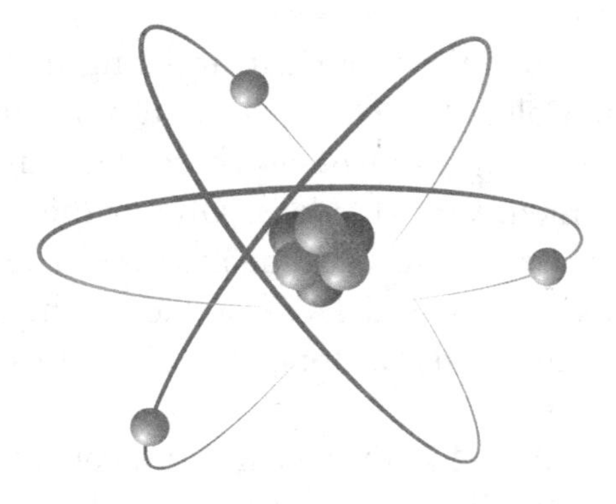

Electricity is a phenomenon that occurs because of these charges. Electric currents occur due to the flow of electrons or of atoms that assume positive or negative charges due to gaining or losing electrons. These charged particles are called ions. A lightning storm demonstrates that electricity occurs naturally all around us, but we also have learned how to generate, harness, and transport electricity for our benefit.

In grade school we learned how Benjamin Franklin, in 1752, used a kite to demonstrate that lightning is a form of electricity. Based on this knowledge, Franklin designed the metallic lightning rod to protect wooden structures from lightning-sparked fires by rapidly conducting the electricity to ground.

MAGNETS

The "magnetic" part of electromagnetic radiation refers to the same type of magnetic fields that emanate from those little pieces of metal that you have sticking to your refrigerator door. Certain materials exhibit magnetic properties (stemming from a particular ordering of the atoms that make up the magnet), enabling them to attract or repel other magnetic objects, or to be attracted to or repelled by other magnets. The needle in a compass is a magnet that points north because it interacts with the magnet that is part of our planet. The space in which the attractive and repulsive forces exert their influence is called the *magnetic field*. And as anyone who has played with two magnets knows, the strength of magnetic fields decreases with distance from the magnet. Because magnetic fields exert influence without making physical contact, physicists refer to magnetism as "action at a distance."

As mentioned above, the earth itself is one giant magnet, with magnetic poles on the north and south ends of the planet. This is why compasses work and why certain species of birds are able to fly such great distances with such accuracy. Human beings also generate magnetic fields (such as those that can be seen in electrocardiograms due to electrical currents in the heart). We measure the strength of magnetic fields in units of *gauss* (G) or *tesla* (T). For some perspective, a typical refrigerator magnet has a magnetic field of 50 G (or 5 millitesla or 5 mT), while your brain emits a magnetic field of approximately .0000001 G.

EMF

Electricity flowing in a current generates magnetic fields. An electrical current (moving charges) in a wire is always accompanied by a magnetic field around the wire. The magnetic fields that result from charge flows are known as *electromagnetic fields* (EMFs) or *electromagnetic radiation* (EMR).

In practical applications involving the effects of emissions from cell phone towers or from cell phone antennas, the strength of EMFs is measured in units of *power density*, which tells us how much power is hitting a particular area. Power density can be measured in watts per square meter (W/m^2) or microwatts per square centimeter ($\mu W/cm^2$), a unit that is 100 times smaller. It is therefore very important to keep track of the units.

While power density measurements inform us how strong a field is, power density does not tell us how much of that power is absorbed by what it comes in contact with, such as a human being. The measure of how much EMF is absorbed by a given area in the field is the *specific absorption rate* (SAR), measured in watts per kilogram (W/kg). Because SAR represents the measurement of radiation absorption at a specific point, it is usually averaged over greater areas, such as the head or body. This is how radiation from cell phones is commonly—though not comprehensively—measured. This approach assumes that radiation is uniformly absorbed throughout body tissue, which is very unlikely.

FREQUENCY

Although the many different types of electromagnetic radiation can be shown as waves, they differ from each other in that they have different *frequencies*, or wavelengths. Frequency is measured in units called *hertz* (Hz). Named after 19th-century German physicist Heinrich Hertz (the first to conclusively prove the existence of electromagnetic waves), Hz quantifies the cycles per second of electricity. It's a measurement that we are all familiar with because it's used to identify the frequencies used by stations in radio broadcasting.

AM radio bands start at around 520 and go up to around 1,610. These are frequencies of EMF (specifically, *radio frequencies*, or RF), where 520 on the AM dial is a signal of EMF radiation vibrating at 520 kilohertz (kHz) and 1,610 is a signal broadcast at a frequency of 1,610 kHz, or 1.61 megahertz (MHz). On the FM dial, you'll find a similar spectrum of stations, which start on your radio around 87.5 MHz and top out near 108 MHz.

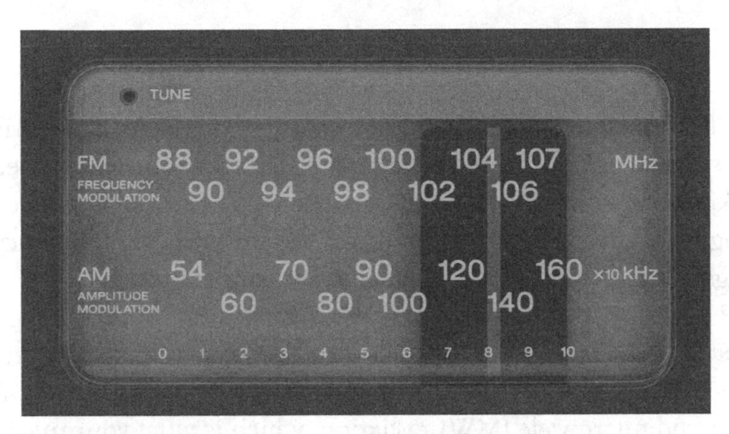

A radio dial showing FM (upper scale in MHz) and AM (lower scale in kHz) frequencies used in radio broadcasting. The dials tune into specific frequencies of electromagnetic radiation that are used to transmit audio signals.

Similarly, the rainbow of colors that make up the range of visible light are defined by individual frequencies. Visible light is a type of EMF—the earliest type of EMF recognized by people. Each color is different because the light vibrates with different frequencies. Red has the lowest frequency of visible light (in the range of 400–484 terahertz, or THz), orange a bit higher (484–508 THz), yellow a bit higher still (508–526 THz), all the way up to violet, which has the highest frequency of any color in the visible light spectrum (668–789 THz). When we say violet has a higher frequency than red, this means that the electromagnetic waves that generate violet vibrate at a faster rate than those waves that make the color red.

We have described waves in terms of their frequency, but one could just as easily describe them in terms of wavelength, as is often done. The product of frequency and wavelength of any wave is equal to the speed at which it advances, and for electromagnetic waves the product is equal to the speed of light, a fundamental constant of nature.

frequency (F) x wavelength (L) = speed of light (C)

This equation also tells us that because the speed of light is a constant, when the frequency increases, the wavelength decreases (and vice versa).

THE EM SPECTRUM

Radio frequency (RF) and visible light are just two ranges of EMF on a broad spectrum of electromagnetic energy. The electromagnetic spectrum (also known as the *EM spectrum*) contains all known frequencies of electromagnetic radiation, from radio waves toward the lower frequency end, through the visible light spectrum, all the way up to gamma rays.

RF is pretty low on the spectrum. Below it, we find *extremely low frequency* (ELF) radiation, such as what is emitted by the power lines and electrical circuits we use to supply electricity to our homes. Above RF, we find microwave (MW) radiation, which is what your microwave oven uses. Above MW is infrared (IR), such as that emitted by motion sensors or remote controls. And, on the other side of visible light, above violet (the highest frequency of color in the visible spectrum), we find ultraviolet radiation, X-ray radiation, and gamma radiation. The EM spectrum is important because different frequencies of EMF radiation are used in many different practical applications.

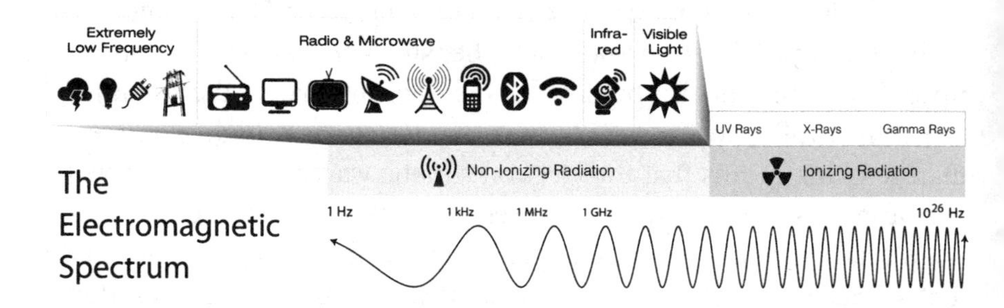

The electromagnetic (EM) spectrum extends from extremely low frequency radiation (pictured here on the far left), up through very high-energy gamma rays (on the far right). Different technologies use EMF radiation from different sections of the EM spectrum.

MAN-MADE AND NATURAL EMF

EMF comes from both natural and man-made sources. Visible light is one type of natural EMF that is generated by the sun. Modern devices like cell phones and WiFi networks generate man-made EMF, as do less novel devices like hair dryers and lightbulbs. As described above, human-made sources of EMF generate different frequencies of EM radiation across the spectrum, generally in the non-ionizing range. Everyday appliances, such as your desk lamp or hair dryer (as well as the power lines that provide electricity to these appliances), generate low-level frequencies of EM radiation in the ELF range. Radio broadcasts are in the RF range. Televisions, cell phones, and their towers emit a higher frequency of EMR called microwave radiation. Humans have been exposed to increasing levels of man-made EMF since the days when electricity was first harnessed. Anything that runs on electricity generates electromagnetic radiation, and our daily lives depend more and more on such products.

Natural EM radiation can hurt you, as anyone who has suffered a nasty sunburn (which is caused by ultraviolet EM radiation) can attest. And some modern conveniences can exacerbate natural EM radiation. For example, when you fly in a plane at an altitude of 20,000 feet, you are exposed to far more *cosmic* EM radiation (the type of radiation that hits earth from extraplanetary sources and which our atmosphere helps reflect) than your body has evolved to handle. This may help explain why flight crews have a higher risk of developing cancer. One of many such studies, for example, indicates that women who have been on flight crews for more than five years have double the normal occurrence of breast cancer.[1]

POWER AND ENERGY

Two terms frequently used in discussing electricity and EMF are *energy* and *power*. In common use, these terms are often interchangeable. In physics, however, these are distinct concepts, and for the purposes of EMF science and safety standards, it is important to understand the difference.

Energy reflects the ability to perform work. The higher the frequency of EMF, the more energy it has (and the more work it can do). So visible light has more energy than RF, which has more energy than ELF (and so on, across the EM spectrum). *Power* (measured in W) reflects the rate at which work can be performed. The higher the power of a given energy, the more work the electricity can perform. So an EMF signal of 300 Hz could be generated with 5 W or 50,000 W, in which case the same signal is being radiated with very different levels of power (the 300 Hz signal generated with 50,000 W of power could travel a much greater distance).

It is a long-held belief that the less energetic EM radiation in the lower end of the EM spectrum is less damaging to the body than higher energy frequencies. ELF is less damaging than RF or MW, and X-rays are more damaging than ELF, RF, and MW. This is why criteria for safety standards are different in each range of the EM spectrum. However, we know that even low-energy EMF can cause bodily damage. A high-powered ELF field can deliver enough current to kill a person (such as a lightning strike or the electric chair), while a person does not even feel a low-powered radio signal (such as the transmission from a nearby baby monitor) that is composed of waves that can be a million times more energetic.

The fact that significant biological responses to EMF can occur across the EM spectrum, shows that the focus on energy levels in the discussion of EMF public health and safety is largely irrelevant. Nevertheless, it has been used to justify significant differences in safety limits for the different groups of EMF known as *ionizing* and *non-ionizing radiation*. Whereas ionizing radiation (the high-energy frequencies of EMF in the part of the EM spectrum above visible light) is widely regarded as hazardous to humans, low-energy non-ionizing radiation (those frequencies of EMF below visible light) has been viewed as much less harmful. This focus on energy has obscured the real biological measures of harmful responses, such as reactions with DNA. The biological studies to be discussed in the coming chapters have shown that such reactions of cells are stimulated at very low energy levels and very low power levels of EM radiation. Ignoring these potentially harmful biological reactions has led to unrealistic safety standards, especially in the non-ionizing ranges.

IONIZING RADIATION

So what is it about ionizing radiation that everyone is so afraid of? As mentioned earlier, all matter is composed of atoms, which have positively charged particles (protons), neutral particles (neutrons), and negatively charged particles (electrons). The protons and neutrons are clumped together in a nucleus, and the electrons move rapidly around the nucleus like planets around the sun. By default, in a stable atom, you find equal numbers of protons and electrons—meaning that the atom is neutral and has no net charge. Which brings us to *ions.*

You may recall from high school physics that an ion is a particular form of an atom (any atom) that has a charge. An ionized atom has a charge because the atom has gained or lost electrons. If the atom loses electrons, that atom is a positively charged ion; if the atom gains electrons, that atom is a negatively charged ion.

Why do ions matter in regards to EMF?

As I've explained, there are different frequencies of electromagnetic radiation. Those frequencies of EM radiation at the top of the spectrum are *ionizing* forms of radiation. Ionizing radiation vibrates at a very high frequency, with a tremendous amount of energy. So much energy, in fact, that when ionizing radiation comes into contact with an atom, it can knock an electron free from its orbit around the nucleus, and the atom becomes a positively charged ion. (The electron can then attach to another stable atom, resulting in a negatively charged ion). In this way, ionizing radiation causes neutral atoms to become charged ions.

Ionizing radiation has long been regarded as extremely dangerous to biological beings—to humans, like you, and all other living creatures. Ionizing radiation causes chemical reactions that, in turn, cause damage to biological systems (like the molecules in your body). So, for example, it has long been acknowledged that prolonged exposure to ultraviolet radiation can lead to skin cancer—this is why you put on sunblock when you go to the beach. Similarly, it is generally acknowledged that you should minimize the number of X-rays to which you are exposed because of their potential to cause damage to your body. And, of course, everyone recognizes the dangers of ionizing EM radiation leaks (in addition to the leaks of radioactive substances) from nuclear fusion reactor failures such as Chernobyl and Fukushima Daiichi.

The scientific community and the public at large recognize the risks of ionizing radiation due to this power to alter the electric charge of atoms and create ions. Forms of non-ionizing EMF, with frequencies lower than that of visible light, do not contain enough energy to force electrons loose—non-ionizing EM radiation cannot cause atoms to become ions. However, as we will see in subsequent chapters, non-ionizing EMF can cause significant chemical changes in important molecules such as DNA.

NON-IONIZING RADIATION

All of the technology and science that are discussed in this book deal with non-ionizing EMF. Cell phones, smart phones, wireless devices, and home cordless phones all generate non-ionizing RF (3 kHz to 300 gigahertz, or GHz) and MW radiation (in the range of 300 MHz to 300 GHz; microwave radiation and radio frequency radiation are often grouped together as RF/MW). Other home appliances and the power lines that feed them generate ELF (from 3 to 300 Hz), which is also non-ionizing.

It has been *assumed* that non-ionizing radiation and the devices that generate it are biologically safe at levels insufficient to heat human tissue. But this is not the case. As I will discuss in the coming chapters, there is a significant body of peer-reviewed, high-quality science that directly and clearly demonstrates that all forms of electromagnetic radiation—including non-ionizing radiation—have observable effects on biological systems. Biological reactions can be affected by exposure to all parts of the spectrum—even in the very low frequency ELF range. *All EMF is bioactive.*

OLD NEWS

As we begin our investigation of the known science linking electromagnetic radiation and negative health outcomes, I wish to emphasize that these ideas are not new. In 1891, Jacques-Arsène d'Arsonval (a French doctor and inventor of the moving-coil galvanometer, which measures electric current) and Nikola Tesla demonstrated effects of electromagnetic radiation on entire biological systems—documenting changes in bodily characteristics including perspiration, respiration, and body weight—resulting from exposure to EMF. And in 1900, V. J. Danilewsky (a Russian clinical investigator) wrote of the effect of "electricity at a distance," arguing that such "long-range" electricity likely affects entire organisms (not just individual biological systems). "Dozens of monographs and thousands of articles devoted to the biological effect" of EMF followed the process of electrification in the United States.[2] You will note that many studies cited in this book date from the 1960s to 1990s.

And yet we find ourselves today, more than 120 years after d'Arsonval's 1891 paper, still debating this fundamental question: whether non-ionizing electromagnetic radiation can cause disease and other negative health effects in humans. As you will see throughout this book, science clearly demonstrates that the answer is yes. Before discussing what science tells us about the biological and health effects of exposure to electromagnetic radiation, it can be useful to examine the scope of the EMF issue. As we'll see in the next chapter, since Thomas Edison's invention of the mass-produced lightbulb, we have been increasing the amount of our exposure to non-ionizing EMF, to the point where today residents of industrialized nations are exposed to multiple frequencies of EMF on a near-continuous basis.

Chapter 3

THE ELECTROMAGNETIC AGE

On January 13, 1976, the *CBS Evening News*, hosted by Walter Cronkite, ended in familiar fashion with the words, "that's the way it is, January 13, 1976." Unusually though, while Cronkite was present and hosting that night, he himself did not say the catch phrase that closed his newscast. Instead, they were uttered by a computer.

That machine (revolutionary for its day, though less powerful than most cell phones today) was created by the then-young inventor Ray Kurzweil, now director of engineering at Google. A few years earlier, Kurzweil Computer Products had become the first company in the world to create what is known as *optical character recognition* (OCR), which gives computers the ability to recognize scanned text. Kurzweil followed that up with text-to-speech synthesis, giving computers the ability to speak text. And the world saw the results that January night on the *CBS Evening News*.

When Kurzweil was born in 1948—the same year that saw the patent of the first transistor with which we now build the integrated circuits in our computers—a human-made machine that could read and speak was just science fiction. But before his 30th birthday, Kurzweil turned it into reality. This is just one example of the many innovations that the world has seen in electrically powered technology since the dawn of the Electromagnetic Age in the late 19th and 20th centuries.

THE LAW OF ACCELERATING RETURNS

Kurzweil is a serial inventor and entrepreneur, though he is perhaps better known to the general public for many of his statements on the history and future of information technology. Kurzweil's forecasts are predicated upon what he terms "the Law of Accelerating Returns."

Many people today are familiar with Moore's Law (named after Intel cofounder Gordon Moore), which predicts that computer processors will double in power every two years. Moore's Law is considered more a framework to establish goals and evaluate performance, than a law per se. However, it has apparently perceived something very fundamental about the way technologies develop, and it has accurately described the exponential growth in the power of computers thus far.

Kurzweil's Law of Accelerating Returns expands on this idea and goes so far as to say that the rate of change in a wide variety of evolutionary systems (including, but not limited to, the growth of technologies) tends to increase exponentially.

While Moore's Law is specifically restricted to the production of semiconductors, Kurzweil's law describes the prevalence and usage of such technology. In other words, just as the power of computer chips has demonstrated exponential growth, so will the number of digital computing devices continue to explode, as will the number of uses to which these devices are applied. Our technology and our consumption of it will continue to expand at an increasing rate.

All products that run on AC power (*alternating current,* on which the power grid runs, unlike *direct current,* DC, on which batteries run) generate the type of EMF that we are concerned with in this book. Similarly, any product that transmits or receives wireless communication signals also emits the type of EMF discussed in this book. Thus, the exponential increases in the rate of technology growth have led to corresponding exponential increases in the amount of human-made EMF to which we are exposed. While cell phones have deservedly received a lot of the attention in the discussion of EMF bioeffects, there are a tremendous number of other EMF-emitting products that surround you in your everyday life.

And it all started with the lightbulb.

A NEW ERA

The lightbulb was such a great idea that it has, itself, become a symbol for bright ideas. Thomas Edison created the first practical lightbulb that was easy to produce on a mass scale nearly 70 years after Humphry Davy presented the first electric lamp to the Royal Society in England. Mass production of lightbulbs created a mass demand for electricity to power them. More than any other single invention, the lightbulb gave birth to a new era in which people would be exposed to human-made electromagnetic radiation in their daily lives.

THE GRID

To help power the market for lightbulbs, Thomas Edison created the first power plant in New York City in 1882. However, this plant produced DC electricity, not the AC electricity that is in use today (and which can be delivered in much greater volume, over much longer distances). The first AC power plant was created four years later. By the turn of the 20th century, most major cities provided AC grid power. By the 1950s, following the US government–led process of *rural electrification*, the grid extended to most inhabited rural parts of the United States.

AC electricity produces electromagnetic radiation in the extremely low frequency (ELF) range, at a frequency of 60 Hz in the United States (50 Hz in Europe and much of the rest of the world). Incandescent bulbs, like the one invented by Edison, use this electricity to heat a filament until it glows and produces light. Unless the bulb is connected to a dimmer, incandescent bulbs run on the same 60 Hz power provided by the grid. (Dimmer switches generate much more electromagnetic radiation due to their manipulation of electrical current.)

Fluorescent bulbs use a different technology that requires more power. So while a 60 watt incandescent bulb can emit 0.3 mG of EMF radiation at a distance of two inches and 0.05 mG at six inches, a 10 watt fluorescent bulb produces 6 mG at two inches and 2 mG at six inches—between 20 and 40 times greater.[1] Compact fluorescent lights (CFLs), which despite the similarity in name, use a different technology from standard fluorescent lighting, also emit much higher levels of electromagnetic radiation than traditional incandescent bulbs. Not much research into CFLs yet exists, though we do know that the EM radiation emitted by

CFLs has a much higher frequency than other bulb technologies. CFLs expose you to EMF more like cordless phones and cell phones than the ELF associated with incandescent bulbs. CFLs also contain mercury, which creates additional problems in terms of breakage and disposal.

Electricity is delivered from the power plant to your home over a network of different types of power lines that can carry different levels of electrical power. Some, known as high-voltage power lines, are like electrical mains, carrying a vast amount of electrical power, between 69 and 765 kilovolts (kV, one thousand volts). In general, the taller the tower supporting the power line, the more powerful the electricity that flows through it. Other lines, such as the distribution lines in neighborhoods that extend power to the transformer that is connected to your home, carry much less power (15 to 30 kV). The EPA, citing the Bonneville Power Administration, estimates ELF exposures from these power lines at different distances in the following chart[2]:

Electric Power Lines					
Types of Transmission Lines	Maximum on Right-of-Way	Distance from lines			
		50'	100'	200'	300'
115 Kilovolts (kV)					
Average usage	30	7	2	0.4	0.2
Peak Usage	63	14	4	0.9	0.4
230 Kilovolts (kV)					
Average usage	58	20	7	1.8	0.8
Peak Usage	118	40	15	3.6	1.6
500 Kilovolts (kV)					
Average Usage	87	29	13	3.2	1.4
Peak Usage	183	65	27	6.7	3

Magnetic field measurements in units of milligauss (mG)
Information courtesy of Bonneville Power Administration.

The grid then extends into your residence, running through your walls across electrical wiring (with much lower levels of power than found running through power lines). As a result, the wiring in your home also emits EMF in the ELF range.

Although the safety issue will be considered later, it is important to realize that actual EMF levels from similar sources can vary. The actual amount of ELF-EMF emissions from power lines and structural wiring will depend on how the cabling and wiring is installed. With both power lines and residential wiring, even seemingly minor changes in design and installation can lead to significant differences in levels of ELF emissions.[3] Certain decisions—such as running the hot and neutral wires together, instead of separately—can significantly reduce EMF emissions. Proximity to the power line and electrical transformer will also affect ELF levels. Some residences have very high levels of EMF, while others do not. Similarly, some areas of your home may have high ambient levels of EMF, while others may be much lower. If you live in an apartment building or work in an office complex, you may also be exposed to ELF emissions from transformers and switching cabinets.[4]

APPLIANCES

With power running to their homes, people started doing much more than just illuminating the night. The 20th century saw multiple generations of consumer products powered by electricity including refrigerators, air conditioners, drills, mixers, blenders, ovens, heaters, coffee makers, food processors, and so on. Like all things that run on electricity, all of these appliances generate electromagnetic radiation. Different appliances generate different levels of EMF radiation. The US Environmental Protection Agency has provided EMF radiation exposure estimates for common household appliances, based on distances from the appliance of four inches and three feet.[5]

DANGER ZONES EMF Levels from Common Sources in Milligauss (mG) Recommened Safety Levels .5 mG–2.5 mG		
SOURCE	up to 4 inches	at 3 feet
Blender	50–220	0.3–3
Clothes Washer	8–200	0.1–4
Coffee Maker	6–29	0.1
Computer	4–20	2–5
Flourescent Lamp	400–4,000	0.1–5
Hair Dryer	60–20,000	0.1–6
Microwave Over	100–500	1–25
Television	5–100	0.1–6
Vacuum Cleaner	230–1,300	3–40
Source: USA Environmental Protection Agency		

EPA estimations of EMF emissions from common household appliances.

It is important to note, as this table does, that most EMF levels drop substantially with distance from the source. This is why the EMF levels at a distance of four inches are so much higher than the values at three feet. For this reason, it is wise to be as far as possible from any household electrical appliances while they are in use. This can, of course, be a challenge with some products, such as hair dryers and electric razors.

Today, AC power reaches virtually all Americans and an estimated 75% of the world's population.

TELEVISION AND RADIO

If you can turn on a radio and hear a radio station, you are being exposed to some level of radio frequency (RF) radiation from radio transmissions. There are multiple bands of radio signals, all within the RF range of the

EM spectrum. AM radio stations in the United States broadcast with frequencies of EMF between 520 kilohertz (kHz) and 1,610 kHz. FM radio stations broadcast at a much higher frequency, between 87.5 megahertz (MHz) and 108.0 MHz. Televisions broadcast EMF at a still higher frequency, in the microwave (MW) range. Standard over-the-air television signals used to broadcast between 300 and 500 MHz. In the United States, digital signals (mandated since 2009) generally broadcast between 54 and 806 MHz, covering and expanding the older analog spectrum.

It is important to remember that while radios and televisions receive wireless RF/MW EMF signals, these devices *also* generate ELF from their AC-connected power supplies. While radios emit very little ELF radiation (the main source of EM exposure from radios is the radio signals themselves, picked up by your radio or stereo), televisions emit much more because of their display technology. Early televisions, using *cathode ray tube* (CRT) technology, literally beamed X-rays from behind the screen directly at the viewer (this is why it was advisable to sit at least six feet away). This is no longer true of the flat-screen displays that rely on a different technology.

Today, television and computer monitor CRTs have been replaced by flat-panel LCD, plasma, and LED displays, which emit much lower levels of EMF. Still, regardless of whether you own an old CRT or a brand-new LCD television, it requires much more power than a radio. Consequently, televisions also emit higher levels of EMF in the ELF frequency than do radios. Television is just one device that results in EMF exposure in both the ELF and MW ranges of the EM spectrum. Another, almost as popular, device is the microwave oven.

MICROWAVE OVENS

At some point in the 1940s, researchers working with television realized that MW radiation could also be used to cook food, giving birth to the microwave oven. The first microwave ovens were developed by the US Navy for use on submarines and were made available to the public in 1947. Today, it is estimated that over 90% of American homes (and many workplaces) have them.[6]

Microwave ovens generate microwave radiation, which is what cooks the food from the inside without the application of heat. While all microwave ovens include protective shielding to minimize leakage of MW radiation, as per Food and Drug Administration (FDA) regulatory guidelines, *microwave ovens are permitted to leak up to 5 mW/cm²*.[7] This leakage applies to brand-new ovens; if you do not service your microwave oven, that leakage will increase over time.

It takes a significant amount of power to generate microwaves sufficient to cook, which is why microwave ovens also emit very high levels of ELF radiation in addition to MW radiation. (You are not protected from this radiation by the oven's shielding—this shielding exists only to suppress the microwaves during cooking.)

CELL PHONES

In the 1950s, researchers discovered how to harness microwave radiation to build telephones. The cordless phone, first invented in 1956 and patented in 1959 by Raymond Phillips, underwent many iterations before it was eventually released to the general public in the 1980s. While many earlier models of cordless phones transmitted 900 MHz microwave radiation, today it is difficult to find a cordless phone that operates under 2.4 gigahertz (GHz), and some even transmit at 5.8 GHz. Base stations for cordless phones also generate ELF radiation from their AC power source, as well as RF/MW radiation to communicate with the remote handset. Cordless phones with DECT, or digital enhanced cordless telecommunications, continually transmit these signals whether or not the phone is actually in use—meaning that such phones continually fill your home with microwave radiation.

Cordless phones are a nice convenience, but they are still tethered to landlines. Severing that physical connection entirely, Martin Cooper at Motorola invented the first cell phone in 1973. The first publicly available cell phone, about the size of a shoe box, was released 10 years later at a cost of approximately 10,000 of today's dollars.

In the years since, the global growth of mobile cellular communication has been simply astounding—and, like the rates of ownership for lightbulbs, televisions, and microwave ovens, is another demonstration of the Law of Accelerating Returns. Cell phone users in the United States increased from 34 million in 1996 to more than 203 million in 2006.[8] By 2009, 83% of American adults had mobile phones—up from 65% just five years earlier.[9] In the rest of the developed world, the picture is much the same. Indeed, rates of mobile phone ownership are much higher in large parts of Europe, where mobile phone ownership is effectively 100%.[10]

Much of the developing world (areas without landline telephone networks), including much of China and Africa, are also seeing tremendous growth in the popularity of mobile devices since they provide regions the ability to become connected without investing in a landline telephone infrastructure. In India, the world's second most populous country, government census data reveals that more citizens have cell phones (53.2%) than toilets (46.9%).[11] Worldwide in 2012, there are estimated to be more than 5.9 billion mobile subscribers, representing approximately 87% of the global population.[12] As Howard Rheingold, professor of digital journalism at Stanford University, has said of mobile phones, "I don't think there is a precedent for something that has spread so quickly around the world to so many individuals."[13]

Martin Cooper holding an early-model Motorola cell phone.
Wixphoto.com for Freerange Stock.

SPECIFIC ABSORPTION RATE

The rate at which the body absorbs energy from radio frequency and microwave radiation is called *specific absorption rate* (SAR). While other measures of EMF, such as power density, tell us the amount of energy in an electromagnetic field, SAR informs us of the amount of energy an exposed subject *absorbs* from the radiation—not the strength of the radiation itself. Accordingly, SAR averages energy over an area of mass and is measured in watts per kilogram (W/kg).

The FCC mandates that every new cell phone must have its SAR determined under specific laboratory conditions and has established 1.6 W/kg as the maximum permissible exposure from these devices. (The private firms that profit from the sale of these devices are those responsible for the testing; the FCC does not test these devices.) As the FCC explains, however, "a single SAR value does not provide sufficient information about the amount of RF exposure under typical usage conditions to reliably compare individual cell phone models."[14]

Why not?

Like all means of measuring and gauging electromagnetic radiation, SAR is a useful tool for scientists working with EMF. However, the manner in which SAR has become a safety standard for cell phones (and similar modern devices) is fundamentally flawed. Stated simply, the SAR of your cell phone does not actually inform how much radiation you are exposed to by using that cell phone.

To establish its SAR, your cell phone was turned on in a laboratory, and the amount of radiation it released was measured at multiple angles and distances. The highest measured exposure becomes the SAR of that cell phone. Thus, the SAR of a cell phone reflects the amount of energy absorbed by a single point of the body, assuming that the phone is used under conditions identical to those under which it was used in the laboratory. Because your phone's SAR rating reflects a single measurement of radiation exposure on a single point of your body, it simply cannot inform how much radiation your body is absorbing from using your phone.

Neither do these limited measurements of radiation exposure consider real-world conditions. For example, the iPhone 4S has a SAR value of 1.18 W/kg delivered to the head,[15] but this assumes that you hold the phone at the precise angle it was held during the test. If you hold the phone at even a slightly different angle, the SAR will be different. By Apple's own account, the iPhone's SAR value is accurate only if one holds the iPhone five-eighths of an inch away from one's head.[16] If you hold the iPhone immediately up against your head, which more accurately reflects everyday usage of the device, the SAR is much higher (neither Apple nor the FCC publish this value).

Further, SAR measurements are taken only when the phone is on and in use. Thus, SAR ratings tell us nothing about how much radiation is absorbed by the body when the phone is on but not in use (perhaps in your pocket, though still in on-going communication with nearby cell towers).

This is why the way in which SAR is utilized for public health concerns over cell phone radiation is of very little utility in understanding how much radiation you will be exposed to from any given cell phone, under any given set of circumstances. For the same reasons, SAR does not correlate with biological responses. SAR is essentially an arbitrary standard devoid of value from a scientific or public health perspective.

And even if SAR did inform us with some accuracy about bodily absorption of cell phone radiation, it would only reflect a single point in time. There are currently no standards for assessing or regulating cumulative exposure to cell phone radiation over an extended period of time.

Please keep all this in mind the next time you see a SAR rating on a new cell phone or wireless device. Some cell phone voice carrier systems (such as Verizon and Sprint, which use CDMA technology) are much worse than others (such as AT&T and T-Mobile, which use GSM technology), especially when the caller is in motion (such as in a moving car or train). CDMA technology is less efficient than GSM at handling the transition of connections between different cell towers, so GSM phones operate at peak power more often. Unless your phone is turned off or in airplane mode, it is in constant communication with these towers. That is, your cell phone continuously (though intermittently) emits RF/MW radiation, even if you are not making a call. SAR refers to the level of radiation only while on a call, and these measurements do not capture ambient emissions, when the phone is idle but on (and possibly in your pocket, immediately against your body).

Thus, the manner in which the FCC determines SAR values for cell phones makes such measurements a virtually useless metric for evaluating your health risk, because such measurements do not accurately reflect your individual exposure. And regulations and public-health guidelines based around permissible or "safe" SAR levels are useless because they fail to consider the wide variety of biological effects shown to occur at low-energy levels.

DATA NETWORKS

As the use of cell phones has spread, so have alternative forms of communication networks reliant upon wireless microwave radiation. WiFi, which operates at 2.4 GHz, broadcasts with a significant amount of power in order to provide on-demand connectivity to anyone in the coverage area (this is why disabling WiFi on your smart phone can extend battery life so significantly). Of course, many homes and offices have WiFi networks. So do many businesses, like hotels, Starbucks, and McDonald's. Indeed, an increasing number of cities have installed citywide WiFi networks, called WiMAX (essentially, long-range WiFi). Whether or not you are running a WiFi device, you are being radiated by these networks. If you are running a WiFi device (such as an iPad), you are being radiated by these networks, as well as the microwave radiation broadcasting from the WiFi card in your device. And if your WiFi device is plugged into a power outlet, that device is also generating ELF.

SMART METERS

Increasingly, power companies around the United States are replacing traditional power meters with wireless *smart meters*, named for their usefulness in helping utility companies monitor and regulate power usage more intelligently. Many have questioned the benefits these smart meters bring to consumers, but there can be no doubt that they are also significant sources of RF/MW radiation. Smart meters use radiation to communicate wirelessly with the utility company (over a much longer distance than a cordless phone, baby monitor, or walkie-talkie), and while they do not communicate continuously, they do so repeatedly, leading to frequent exposures for those nearby (or on the other side of the wall from) the smart meter or where many meters are grouped in large apartment blocs. Some smart meter systems are networks where one meter collects from several other meters and transmits all the information.

OTHER WIRELESS DEVICES

As wireless communication technology continues to decrease in cost and increase in power (per Moore's Law and the Law of Accelerating Returns), so do the variety of devices in which we find wireless communication features. Remote controls generally rely on infrared (IR) electromagnetic radiation to carry their signal. Video-game controllers use different types of EMF, depending on the model. Baby monitors are similar to cordless phones and are continually transmitting while turned on. Similarly, walkie-talkies broadcast using RF frequencies and CB radio (or citizen-band radio) transmits RF EMF at a frequency of 27 MHz. Progressive now offers a discount for auto insurance customers who sign up for their Snapshot program in which a device is placed in the car and repeatedly communicates driving behavior to Progressive using a cell phone signal. There are thermometers that run wirelessly on RF electromagnetic radiation. There are also wireless electric dog fences and wireless pest repellents that transmit RF EMF. RF and MW radiation sources are now found in almost every room of people's homes, as well as in the yard.

COMMUNICATION ANTENNAS

While the devices with which we surround ourselves and the grid that powers them are both significant sources of electromagnetic radiation, there is another major source—one that is frequently hidden and easily ignored: the network of powerful antennas that we use to make our cell phones, televisions, and radios work.

Building a functional cell phone prototype and selling a finished, consumer-ready cell phone to the public were two very different accomplishments. Indeed, turning inventor Martin Cooper's 1973 prototype into Motorola's 1983 DynaTAC 8000X involved a staggering number of engineering breakthroughs, not the least of which involved producing an infrastructure. Mobile cell phones require a network of broadcast antennas, or *cell towers*, in order to make and receive telephone calls wirelessly. Cooper made the first cell phone call in 1973 using two antennas that Motorola built specifically for the project. For the public to use this new technology, however, many, many more would have to be erected. This would prove to be such a costly and time-consuming process that many doubted cell phones would ever become a viable business.

These naysayers were, of course, proven quite wrong, as the globe is now dotted with these cell network antennas, often mounted on towers (typically over 200 feet tall), which can house multiple cell antennas each. Antennas are also positioned on top of offices, hospitals, apartment buildings, churches, light poles, and signs. In 1985, just two years after Motorola's release of the first public cell phones, the United States saw approximately 900 cellular towers erected (making for some pretty poor network performance). By 2005, that number had grown to 175,725.[17] As of July 1, 2012, the website http://AntennaSearch.com noted 480,058 cell phone towers and 1,535,883 cell phone antennas in the United States alone.[18] Accurate global statistics on towers and antennas can be difficult to find, but there are millions worldwide, their numbers growing apace with the increase in cell phone ownership and the increasing data-flow requirements that we, as users, are placing on these networks.

Each of these towers and antennas continually sends and receives EMF radiation. This is how you are able to make and receive cell phone calls at any time, as long as you are in service range of your cell provider's network. Regardless of whether your phone is on or if you even own a cell phone at all, you are exposed to radiation from these towers, which broadcast EMF into the environment.

An antenna is camouflaged on the tower of this church in Sopot, Poland. See the close-up on the left. Photo by Piotr Plecke.

Increasingly, we find more and more "hidden" antennas, so-called because they are designed to blend into the environment. These are even more difficult to avoid, as they are designed to be hidden. I encourage all of you to visit http://AntennaSearch.com; input the address of your home, office, or child's school; and see how many antennas exist within a four-mile radius. The results may well surprise you. If you have a cell signal, you are being exposed to radiation from at least one tower (and likely more than that).

Like cell phones, televisions and radios can function properly only when supported by a network of powerful communication antennas that relay RF/MW communication signals. These are usually mounted on towers or tall buildings, and can transmit 50 kW (that's 50,000 watts) of energy. That is a tremendous amount of energy, and it explains why exposures are so high in areas close to radio and television broadcast antennas.

DIRTY ELECTRICITY

The alternating current power that provides electricity throughout the outlets and sockets in your home creates a landing place for what the utility industry has dubbed *dirty electricity*. In contrast to *clean* 60 Hz AC electricity, which is produced by utilities and delivered over the nation's power grid, dirty electricity refers to all the other EMF noise (frequencies in addition to 60 Hz) that is picked up by power lines. In essence, the power grid is one giant radio antenna. Individuals are exposed to this dirty electricity when it is carried into residential electrical wiring. Dr. Sam Milham, a physician and epidemiologist, and author of the book *Dirty Electricity: Electrification and the Diseases of Civilization*, believes that exposure to this type of EMF noise is the source of significant health problems. He has documented the basis for this idea in numerous case studies (some of which are discussed later in this book).

PROFESSIONAL EXPOSURE

Certain careers expose their workers to a higher level of EMF than the general public. Not surprisingly, this includes electricians and electrical utility workers (both line workers and employees at power plants), as well as welders, seamstresses, rail-line workers (as many railways are run on high-powered electrical currents that run in the tracks), doctors, dentists, air traffic controllers, airplane crews, communications operators, military radar operators, and construction workers who frequently use power tools.

ON EVEN WHEN OFF

As already mentioned, cell phones and many cordless phones transmit microwave radiation whether or not they are in use (as long as they are powered on). Many wireless devices are continually broadcasting and searching for new networks if they are not already connected to one. Similarly, it is impossible to fully power-down most televisions without unplugging them—there is always a light on the front of the display that indicates that the television is, in fact, needlessly consuming power and generating ELF radiation. Increasingly, all of the devices we surround ourselves with are powered on to some degree—even when they are off. I'm not clear why this has become a standard feature of consumer products, but of course, in addition to being a waste of energy, this type of design leads to even higher levels of electromagnetic exposure and ambient EMF released into the environment.

THE ELECTROMAGNETIC AGE

The history of the 20th century is inseparable from the technological developments involving the use of electromagnetic radiation. The preindustrial world featured no non-ionizing electromagnetic radiation, outside of sunlight (and low levels of radiation from other cosmic sources), lightning, and geomagnetic forces. As humans came to understand the power of EMF, we have harnessed it for an exponentially increasing number of applications. As a result, unlike other known environmental pollutants such as PCBs, CFCs, or lead, EMF isn't a by-product of civilization. To the contrary, EMF science and our ability to harness it are the very cornerstone of modern society. This is why I believe the time we are living in can be termed the *Electromagnetic Age*. Just as tools derived from bronze and copper defined the Bronze Age, and iron technology defined the Iron Age, so does EMF science and technology define today's world.

The early accomplishments of the Electromagnetic Age have been truly remarkable. We sent a man to the moon. We split the atom. We decoded the human genome. We have virtually eliminated certain previously catastrophic diseases from human populations. We have created the Internet and an amazing array of devices that enable us to communicate with virtually everyone on the planet at the speed of light. The power and technology that generate EMF have fueled staggering, previously unimaginable achievements.

Throughout the Electromagnetic Age, applications of electromagnetic technology have been moving up the EM spectrum, increasing in frequency and energy. The start of the 20th century saw the rollout of

ELF-generating AC power. Then radio moved up into the RF range. Television, radar, cell phones, and WiFi networks utilize even higher-frequency microwave radiation. As a result, we find ourselves early in the 21st century increasingly bombarded by more non-ionizing electromagnetic radiation, coming from more sources, across more of the EM spectrum. It is increasingly difficult to find any escape from the ubiquitous *electrosmog*.

LOOKING AHEAD

My parents were born into a world without human-made EMF radiation in their environment. That, obviously, had changed by the time I was born, as I enjoyed modern benefits such as refrigerators and lightbulbs. AC-powered technology (that generates ELF EMF radiation) and wireless communications technology (that generates RF/MW EMF radiation) exploded at an exponential pace in the 20th century. It is clear that this trend is set to continue—that the EMF-generating devices on which we've all come to rely will continue to multiply at an increasing rate.

Using humanity's rate of progress in 2001 as a benchmark, Kurzweil explains that the 20th century saw 25 years of progress. Applying the same 2001 benchmark, Kurzweil predicts that the 21st century will see 20,000 years of progress. Accordingly, we can expect to be surrounded by a continually expanding amount of electromagnetic radiation from the tools and devices that define modern civilization.

With each passing year, each of us can expect to be exposed to more electromagnetic radiation from an increasing variety of sources. The rate of proliferation of EMF-generating technologies is exponential, and there are now so many different sources that many often go almost entirely unnoticed.

As we will see later in the book, the types of EMF exposures that these technologies can lead to are associated with many negative health outcomes, including cancer, Alzheimer's, and infertility. Before exploring how EMF is linked with disease and ailments, we will first investigate what science demonstrates about the biological effects of EMF exposure. In other words, how exposure to non-ionizing electromagnetic radiation alters your DNA and cellular function, paving the way for the onset of disease.

Chapter 4

EMF DAMAGES DNA

One evening in 1953, a research scientist at the University of Cambridge in Britain walked into the Eagle, a well-known pub, and announced that he had "found the secret of life."[1] Turns out that he was right.

That man was Francis Crick, and he and his collaborator, James Watson, were about to publish one of the most significant and influential discoveries in modern science: a molecular model of deoxyribonucleic acid, or DNA. At last, the mysterious structure with its vital clues about the mechanism underlying heredity was revealed. The world would finally begin to understand the mechanism of inheritance—how living beings are brought into existence.

In the years since the Watson and Crick discovery, we've learned that DNA is an intricate and versatile molecule. Indeed, while the structure of DNA was discovered in 1953, it took another 47 years—and the invention of the supercomputer—to begin to decode the genome—the full set of genetic information represented inside of your DNA.

DNA is not only intricate, but delicate, and susceptible to damage—the type of damage that is believed to cause cancer and other serious diseases. Scientists, doctors, and researchers have long accepted that ionizing radiation—such as the ultraviolet rays that accompany sunlight, or the X-rays that you are exposed to in your doctor's or dentist's office—can harm and destroy DNA. It has been assumed, however, that non-ionizing radiation from power lines, television broadcasting, and cell phones did not harm DNA. For instance, in 2002, Dr. Robert L. Park of the American Physical Society, stated:

> All known cancer-inducing agents . . . act by breaking chemical bonds, producing mutant strands of DNA. Not

> until the ultraviolet region of the electromagnetic spectrum is reached . . . do photons have sufficient energy to break chemical bonds. Microwave photons heat tissue, but they do not come close to the energy needed to break chemical bonds, no matter how intense the radiation.[2]

Powerful statements from prestigious organizations and respected academics, such as the one above, aim to assure the public that EMF is not a health hazard. You may well wonder how I can claim that dangers do exist. Well, the answer is that Dr. Park may know physics, but he is grossly uninformed about biology.

Biological science clearly demonstrates that all frequencies of EMF—including the non-ionizing radiation created by your cell phones, laptops, tablets, and the WiFi antennas in them—can react with and damage DNA. The DNA damage can then lead to cell death or remain as a mutation, which can lead directly to serious diseases, including cancer. This chapter will explain how this happens.

CELLS

All plants, animals, insects—all the organisms that you see around you that are alive—are made of cells. Their existence was first noted in 1665 by British philosopher and researcher Robert Hooke. Hooke, who was the curator of experiments (akin to the director of research and development) for the Royal Society in London, was an early pioneer of the use of microscopes in the field of biology. It was while using a microscope to examine a piece of cork that he made his great discovery, becoming the first person to see and publicly identify cells. (The microscope he used for this research is in the collection of the National Museum of Health and Medicine in Silver Spring, MD.)

It goes without saying that Hooke's discovery of the cell was significant. At the time, however, Hooke did not appreciate what it was he had found, and it was not until over 170 years later, around 1839, when *cell theory* finally emerged.

Cell theory is based on a few core principles:

- All living things are made of at least one cell.
- All cells come from other cells.
- Cells perform functions vital for the survival of the organism.
- Cells contain hereditary information.

That last point on heredity proved to be a tricky one. While scientists learned much about cells and their function in the intervening century, it wasn't until the 1950s that we understood what role DNA played in heredity. The three main aspects of DNA that will help us understand the harmful effect of EMF include the electrical properties of DNA, the replication process of DNA, and DNA's role in coping with environmental stresses at the cellular level.

ELECTRICITY

DNA—with its two strands of genetic information, intertwined in a beautifully symmetrical twisting ladder—is among nature's most impressive formations, a shape we call the *double helix*. Though the double-helix structure was noted in the original paper on the model of DNA, the term "double helix" was popularized much later, in the title of James Watson's 1968 book, *The Double Helix: A Personal Account of the Discovery of the Structure of DNA*.

DNA is incredibly efficient, too. Inside of every cell in your body, you have a staggering six feet of DNA, coiled up in twists and folds to fit inside of the nucleus at the center of the cell—it's an amazing feat of engineering with some very interesting electrical characteristics.

The intertwined strands of DNA are connected by rungs of molecules called *nucleotides* (sometimes also called *bases*). Each rung of your DNA is composed of two nucleotides, one from each strand, bonded in pairs. These nucleotides are held together by hydrogen bonds, where a single hydrogen atom, shared by two nucleotides, acts as the glue. The presence of so many nucleotides, connected with hydrogen bonds, results in a strong attraction between the two strands.

These hydrogen-bonded nucleotides are relatively flat molecules with electrons on both surfaces. Because the rungs of the DNA ladder are very close, the electrons form a continuous layer (often described as an *electron cloud*) that is able to conduct an electron current along the DNA chain. This makes it easy for electrons to be conducted (as in a wire) along the nucleotides that form your DNA rungs, a phenomenon called *electron transfer*. If an electron is released to the DNA rung by an oxidizing agent, that negative charge will flow through the nucleotides. Jacqueline K. Barton and her group at the California Institute of Technology have demonstrated long-range electron transfer in DNA and how this ability can vary with the composition of the base pairs.[3]

In other words, DNA conducts electricity.

DNA is such an efficient conductor of electricity, in fact, that it is a common building material in molecular electronics, or biological nanotechnology. Researchers who build extremely tiny machines out of living matter use DNA as one of their construction materials, precisely because it conducts electricity so well.

FRACTAL ANTENNA

Another fascinating electrical trait of DNA stems from its compact shape. One of the ways in which the six-foot-long DNA molecule is able to fit so efficiently in a space as tiny as the nucleus of a single cell is by packing itself into a tightly coiled fractal pattern. A fractal is a shape that displays self-similarity, where each part of the shape looks like the entire shape. No matter how far you zoom in or out, the shape looks the same. DNA is a fractal because the smaller coils are themselves coiled into larger coils, a shape known as a *coiled coil*.

It turns out that the coiled-coil structure and electrical conductivity seen in DNA are the two key characteristics of what we call fractal antennas. The coiled-coil structure of fractal antennas maximizes the length of the antenna, while minimizing its overall size. As a result, fractal antennas are both very long and exceedingly compact. This design can boost cell phone signal strength or radio reception, and amplify a wide range of electromagnetic frequencies. And DNA not only looks like a fractal antenna, but acts like one as well.

By definition, a fractal antenna can pick up and react to a wide range of frequencies of EMF, which means that many frequencies of EMF in the environment can and do react with your DNA. This is why DNA is very sensitive to electromagnetic radiation—notably more sensitive to EMF than other large molecules (such as proteins) in your body.

DNA REPLICATION

DNA is closely linked in people's minds with inheritance—why you look like your parents or why dogs give birth only to puppies. But if that were the end of DNA's importance, it would have little impact on your life after your birth.

Far more dynamically, DNA plays a vital and ongoing role throughout your life. Cells inside your body continually die, and new ones are constantly created to replace them. New cells are needed to sustain growth. This is how newborns grow into infants, then into children, and then into adults—by increasing the number of cells in their bodies and replacing those cells that die off. In children, this process is particularly dramatic. But even as adults, our bodies continuously produce new cells. For instance, on average we replace all the cells in our stomach lining every four days. Other multicellular structures and organs have slower cycles, but cell reproduction is going on in our bodies all the time.

For a new cell to be created, the DNA in a cell (referred to as a *parent* cell) has to be copied. This copying, or *replication*, process in DNA is truly amazing. Every time a human cell divides, its DNA replicates, copying and transmitting the exact same genetic data to the new cells. Given the trillions of cells in our bodies, the number of replications occurring on a daily basis is mind-boggling.

But cells do not do this perfectly.

MISTAKES

The objective of DNA replication is to create exact copies of the original DNA. However, given the immense scope of the DNA replication process, it's to be expected that mistakes will happen—and they do. It is estimated that DNA makes replication mistakes 0.001% of the time. That may sound low (imagine a baseball pitcher who allowed only one hit for every 100,000 batters), but given the amount of DNA in each cell, there are approximately 120,000 mistakes in the DNA each time one of the cells in your body divides.

One of the most common types of error is termed a *strand break*, when a DNA chain breaks apart. When the break is in one strand of the DNA's double helix, it is termed a *single-strand break*. When the break occurs in both strands, it is called a *double-strand break*.

FIXING MISTAKES

Fortunately, the cells in your body also contain tiny quality-assurance testers. These testing mechanisms validate the DNA copies, checking for mistakes. And in many cases, your cells are able to fix the mistakes. So while strand breaks occur all the time, the cell has repair mechanisms to correct many of these breaks. Indeed, studies show that the better your cells are at repairing this damage, the longer you'll live.

Still, no matter how good your cells are at making repairs, there is a limit to what can be fixed. So when mistakes such as these strand breaks become too numerous (and that level will be different for each person), the cell's repair mechanisms cannot cope and the damage remains. The DNA in the new cell has mutated from the original.

Often, mutated cells cannot function properly. When this occurs, the cell activates a process named *apoptosis*, or *programmed cell death*, to kill and remove the cell. This is an optimal outcome, because once the damaged cell is dead, it cannot harm the body or pass on its defective genes.

Sometimes, however, the damaged cell with the mutated DNA survives and replicates, becoming a permanent genetic mutation in the body. Sometimes such genetic mutations are harmless, or at least the damage is irrelevant to the cell's operation. All of your DNA is in every cell of your body, but not all cells need all genes. For instance, the cells in your brain express different genes than the cells in your skin, and so on. Other times, however, the damage mutates the DNA (and its offspring) in a harmful way. This is the process through which diseases like cancer are believed to develop.

A variety of forces, both internal (your family's genes) and external (such as exposure to pollutants), affects the rate at which DNA damage occurs. EMF is one of these forces.

EMF BREAKS DNA STRANDS

One of the most important series of studies on the question of DNA damage from exposure to non-ionizing electromagnetic radiation was performed by Drs. Henry Lai and Narendra Singh starting in 1994 and running through 1998. Lai and Singh, working at the University of Washington, wanted to answer a simple question: Does non-ionizing radiation damage DNA? To make the results more applicable to daily life, Lai and Singh decided to use levels of EMF radiation considered "safe" by government standards.

Professor Henry Lai, University of Washington
Columns Magazine, March 2005

Their results showed that even exposures of only two hours increased the frequency of DNA strand breaks in the brain cells of living rats. Lai and Singh then performed similar experiments with lower frequencies of EMF. For instance, they exposed their subjects to EMF in the range you would find in your average desk lamp. Once again, they found increased occurrences of strand breaks. Lai and Singh's research, demonstrating DNA strand breaks following exposure to non-ionizing EMF with field strengths of 0.25 or 0.5 millitesla (mT) has been replicated in other laboratories, and it clearly demonstrates that EMF can damage DNA even at low EMF-exposure levels.[4] The levels of radiation at which Lai and Singh demonstrate this damage are well below the limits set by the current safety standards for technologies like cell phones, WiFi networks, and microwave ovens.

Even more disturbingly, Lai and Singh found that the DNA in the rat brains continued to break down for hours after exposure ended.[5] This suggests that the exposure not only causes immediate damage, but also unleashes a chain of processes that continue to produce damage well after the exposure itself.

Many other studies have found similar *genotoxic* effects (effects that are poisonous to DNA) resulting from EMF exposure. As Dr. George Carlo and Martin Schram explain in their book *Cell Phones: Invisible Hazards in the Wireless Age*, multiple studies have demonstrated increased rates of micronuclei in the body following exposure to RF/MW radiation (such as that emitted by cell phones). A micronucleus is a fragment of DNA with no known purpose, a by-product of errors that occur during cell division. The presence of micronuclei indicates a type of DNA damage so strongly associated with cancer that doctors test for them as a means of diagnosing cancer.

In 2009, Hugo W. Rüdiger, a professor at the Medical University of Vienna, released a study analyzing the results of 101 different published articles on the effects of low-frequency EMF on DNA. His review, published in the peer-reviewed journal *Pathophysiology* in August 2009, found that "of these, 49 report a genotoxic effect and 42 do not. In addition, 8 studies failed to detect an influence on the genetic material, but showed that RF-EMF enhanced the genotoxic action of other chemical or physical agents." He concluded that "there is ample evidence that RF-EMF [low-energy radio frequency electromagnetic fields] can alter the genetic material of exposed cells."[6]

PROTEIN SYNTHESIS

Another area in which low-frequency, non-ionizing radiation has been proven to harm DNA function is in your cell's production of protein. Protein inside your cells is just like the protein your doctor tells you you should eat more of. In fact, it's precisely because your cells need protein to function that you should be eating a healthy amount of the right types of protein. Protein is required for *all* the functions that your cells perform.

The science of proteins extends back to the 18th century, and since that time, an increasing number of proteins have been discovered and classified. Many of the body's proteins deal with what might be considered basic housekeeping functions of life such as developing muscle to enable us to move and enzymes to enable us to digest food. Another class of proteins is antibodies, first identified in the 1890s, which your body's cells create when you are under attack by foreign organisms (like a cold). The most recently discovered class of proteins is stress proteins, which are stimulated by potentially harmful environmental agents, including EMF.[7]

Fortunately, you don't need to eat all of the exact proteins that your body needs. And even if you did, they would not be in the form that your body requires. That's because your cells convert the proteins you eat into the types of proteins that it needs to function. When you eat protein, your body breaks that protein down into its constituent amino acids. Your cells then take these amino acids and build new proteins. Replacements are needed for damaged proteins, and new proteins are needed for new cells that are formed when cells divide. Because so many proteins have to be made, protein synthesis is happening in your body all the time.

DNA is central to protein synthesis. Human DNA has the ability to create about 25,000 different kinds of proteins; with those, your body can work to create an estimated 2,000,000 different types of protein that your body needs to function properly. Some proteins are always present (such as those that aid in the process of food digestion), and some proteins are created by your body on demand (such as antibodies that aid in the defense against viruses).

STRESS

While your body has a vast range of proteins that serves millions of functions, one type of protein is particularly relevant to the EMF issue: the proteins that help your cells cope with damage from environmental stress.

In the 1960s, Ferruccio Ritossa, an Italian scientist, made an unusual discovery in the chromosomes of a fruit fly. Over the course of an experiment, the flies were accidentally exposed to a temperature increase of a few degrees. Ritossa noticed that when this happened their chromosomes became enlarged at particular sites. It would be another 15 years before the significance of the finding was realized. When it was, the *heat shock response,* as it came to be termed, was found to occur in both animals and plants—including yeast. After extensive research in laboratories around the world, it was found that the heat shock response was the first stage in the synthesis of a special class of proteins called *heat shock proteins.*

Heat shock proteins repair other proteins. They serve a defensive role, defending cells against the ill effects of increases in temperature that could otherwise prove fatal. What's more, they strengthen the cells to be more resilient to temperature increases in the future. Just like lifting weights today makes you stronger tomorrow, these heat shock proteins make your cells better able to cope with subsequent stresses of increased temperature. This greater resistance to the stress of heat shock is called *thermotolerance.*

Since the initial discovery of the heat shock response and heat shock proteins, we have discovered that cells produce similar proteins to cope with many different types of stress—not just temperature. And so, these proteins came to be collectively referred to as *stress proteins,* which are involved in the *cellular stress response*—the process by which individual cells cope with stress. (Because of the way these proteins were discovered, they are still designated with an "hsp" to show that it is a heat shock protein, and a number that is related to size.)

What stresses a cell? A variety of forces, as it turns out. The presence of heavy metals, changes in acidity, alcohol, viral infections, ultraviolet light, and low-oxygen conditions—all of these can damage your cells in the same way that an increase in temperature can. So, there are a variety of conditions that lead to the stress response—a rise in temperature, or thermal stress, is just one of many environmental factors. And on the whole, the cellular stress response has been very effective in helping cells cope with environmental stresses.

EMF AND THE STRESS RESPONSE

EMF—even at low-energy, non-ionizing frequencies—is among the environmental factors that stimulate the stress response in the cells of our bodies. Dr. Goodman and I demonstrated the relationship between EMF and the cellular stress response in research that we performed in 1994. We released a comprehensive review of the studies on EMF stimulation of stress protein synthesis in 1998, and an update in 2009 in the special EMF issue (volume 16) of the journal *Pathophysiology*.

Our initial studies found that when human cells are exposed to radiation in the extremely-low-frequency range (ELF similar to that from power lines), the stress response is triggered and cells begin to create stress proteins within five minutes. EMF appears to stimulate stress protein synthesis in much the same way that the natural electric fields that transmit signals in your nerves lead to the creation of proteins in your muscles.

We later repeated our experiment using EMF in the frequency range emitted by cell phones, and we found the same effect. These findings have been subsequently replicated in multiple experiments by researchers around the world.[8]

So we know that the presence of stress proteins is an indication that the cell has come into contact with something that it reacts to as harmful. And we know that EMF triggers the presence of stress proteins. If our bodies generate stress response proteins for EMF, doesn't that mean that our bodies are coping with whatever damage electromagnetic radiation might otherwise cause?

Yes and no.

As we've seen, the cellular stress-response can be extraordinarily useful because it allows the organism to adapt to and overcome problems. The defensive value of the stress proteins is undeniable, and it is generally accepted that the short-term results from the generation of stress proteins in your cells is almost always beneficial. Stress proteins build your body's defenses against damaging forces like increases in temperature and reductions in the oxygen supply that could otherwise prove life threatening. Because the body generates stress proteins to strengthen cellular resistance to EM radiation, you are well equipped to handle limited exposure to EMF-generating devices such as cell phones. However, there are limitations.

ELECTROMAGNETIC TOLERANCE

The long-term effects, however, are a different matter. Scientists have found that prolonged exposure to EMF (which, again, in the short term encourages the generation of stress proteins) has the opposite effect—extended exposure to EMF *reduces* the ability of your cells to produce stress proteins. Dr. Goodman and I first showed in 1996 that there was a decreased response when EMF stimuli are repeated.[9] A reduced stress response is similar to the thermotolerance that results from prolonged exposure to heat shock. The evidence shows that extended exposure to EMF begins breaking down your DNA's cellular stress response. Given the rise of wireless and other electronic technologies, more people are increasingly subject to prolonged EMF exposures and potentially developing an *electromagnetic tolerance* at the cellular level.

Drs. A. DiCarlo, J.M. Farrell and T. Litovitz at Catholic University of America in Washington, DC, observed similar results in an experiment performed on chicken embryos.[10] In those eggs exposed to ELF radiation of 8 μT (such as that emitted by power lines) for 30 or 60 minutes at a time, twice a day for four days, production of hsp70 (heat shock protein 70, created in response to oxygen deprivation) declined. The same response was noted in eggs exposed to RF radiation (such as that emitted by cell phones) of 3.5 $\mu W/cm^2$ (microwatts) for 30 or 60 minutes, once a day, for four days. The researchers noted that these eggs produced 27% less hsp70 following these exposures and had correspondingly reduced cytoprotection (the ability to fend off cell damage).[11] Similar experiments have been carried out with short, repeated exposures (in contrast to extended exposures). There, too, the rate of stress protein synthesis is reduced with each repetition. These experiments clearly demonstrate EMF tolerance.

Long-term exposure to EMF (either prolonged or repeated or both) reduces your body's resilience to stressful forces in the environment. Our species did not evolve with all of these external electromagnetic forces continuously impacting our bodies. Today, your body's cellular stress response is being called into action in a way it is not prepared for. With each additional moment you are exposed to electromagnetic fields, your cells (the basic building blocks of your body) are more susceptible to damage from other harmful forces in the environment, such as the sun's ultraviolet rays. EMF reduces your cells' ability to respond to many types of environmental damage.

And the effect is cumulative, across a lifetime of exposures.

Long- and short-term exposure to electromagnetic radiation can harm DNA in your body, leading to cell death and cell mutation. These effects are seen across the EM spectrum—not just from ionizing radiation like ultraviolet and X-ray, but also from non-ionizing radiation including cell phone MW transmissions and even the extremely low frequency EMF from power lines.

These are some of the most important effects of EMF exposure at the cellular level. The type of cellular damage caused by EMF is similar to that caused by aging. The residual errors and genetic mutations accumulate, leading to malfunction and disease. There has been a steady rise in EMF radiation in the environment, and DNA damage is now occurring more frequently and earlier in life because of the many ways modern technological forces permeate our lives.

When considering the risks of the EMF issue, people are generally interested in the *health effects*—outcomes such as cancer and other diseases. The types of cell damage described in this chapter are among the known *biological effects* shown to result from non-ionizing EMF exposure. The biological effects are mechanisms and indicators of the health effects that will be described in coming chapters. The reduction in cells' ability to invoke the stress response leaves us more susceptible to disease. DNA mutation is a process by which cancers are widely believed to form, and the DNA damage indicated by the presence of micronuclei is considered to be a strongly accurate indicator of cancer.

Now that we understand some of the biological effects of EMF exposure and that these are associated with the generation of disease in the human body, let's examine what science can tell us about the link between exposure to non-ionizing electromagnetic radiation and cancer.

Chapter 5

EMF AND CANCER

In the middle of the 19th century, the SoHo district in London's West End was not the trendy fashion, retail, and dining area that it is today. Once a pastoral setting of fields and farms, by the 1850s, "Soho had become an unsanitary place of cow-sheds, animal droppings, slaughterhouses, grease-boiling dens and primitive, decaying sewers."[1] London's relatively new sewage system had not yet reached this part of town, and many cesspools were overflowing into basements and cellars.

Paul-Gustave Doré, *Over London–by Rail*, c. 1870, engraving. From *London: A Pilgrimage* by Paul-Gustave Doré and Blanchard Jerrold (1872).

Today we understand that such conditions breed germs and infections, but this was decades before germ theory became widely accepted. At the time, though none of the residents knew it, SoHo was a perfect breeding ground for killer bacteria.

THE 1854 CHOLERA OUTBREAK

On August 31, 1854, residents of SoHo started dying. Only three days later, 127 people had already succumbed to the cholera outbreak.[2] The death toll would reach over 600 before it ended. What was the cause? Why and how were all of these people contracting cholera?

A prominent and well-respected doctor named John Snow wanted to find out. By this time, Snow was a member of the Royal College of Physicians and a government advisor who had spent several years studying cholera. Snow was a skeptic of the widely accepted miasma theory, which posited that diseases were caused by coming into contact with polluted, or "bad," air. Instead, Snow had begun to believe that small biological organisms called germs were the cause of diseases like cholera.[3]

Based on his prior research, he suspected water was the most likely source. So Snow did something simple and obvious—he went to SoHo and began talking to the remaining residents, asking them questions about life in the area and about their water consumption habits. He then compiled and analyzed the results.

From those interviews, Snow deduced that the source of the cholera outbreak was probably the water pump located on Broad Street in the center of the neighborhood. He approached the town council with his evidence, the council shut down the pump on September 8, and the outbreak ended.

Once the outbreak was contained, Snow wanted to ensure that each case of cholera during the outbreak could be traced back to the Broad Street pump. So he plotted all of the cholera deaths on a map of the neighborhood. For all but 10 of the victims, the Broad Street pump was the closest available water source, and of those 10, Snow discovered through interviews how each had made occasional use of the Broad Street pump. In Snow's words, "the result of the inquiry, then, is that there has been no particular outbreak or prevalence of cholera in this part of London except among the persons who were in the habit of drinking the water of the above-mentioned pump well."[4]

Once the pump was dismantled, Snow took it back to his laboratory for further investigation, but he was unable to identify any chemical or biological causes for the cholera. (The cause was eventually identified as a breach in a cesspool located only feet away from the Broad Street pump containing the diaper of an infant who had died from cholera.) Still, without any supporting laboratory evidence—without any absolute "proof"—Snow was able to determine that the SoHo residents who contracted cholera did so by drinking water from the Broad Street pump. He did this solely by assembling, analyzing, and drawing scientifically based conclusions from a large amount of accurate data.

Dr. John Snow's map of cholera fatalities (black dots) during the 1854 London outbreak, showing the locations of pumps (X), with the Broad Street pump in the center.

EPIDEMIOLOGY

Over the course of his career-long effort to identify the source of cholera and make the environment healthier for human beings, Snow pioneered many scientific research and analytical techniques, solving multiple problems that were impossible to answer in a laboratory. In so doing, Snow developed a whole new branch of science—epidemiology.

Epidemiology is the science of studying patterns, causes, and effects of health and disease in a given population. It enables us to research and answer questions that would otherwise be immoral or impossible to research in a laboratory. Today, epidemiology is a critical component of both modern science and public health, providing key insights on a huge number of important questions.

Laboratory studies (or experimental science), which is able to isolate variables, can sometimes prove causation—that one thing causes another. If we expose DNA to EMF, we see an increase in strand breaks. This indicates that EMF is a cause of DNA strand breaks. If others repeat the experiment and get the same results, we have repeatable scientific evidence that EMF causes strand-break damage in DNA. This is what we mean by scientific proof.

Because it works differently than traditional laboratory science, epidemiology cannot prove causal relationships. Instead, epidemiology relies on the analysis of statistical data to establish correlations, or the relationship between two or more sets of data. For example, by examining the drinking habits of those infected during the 1854 outbreak, Snow demonstrated that there was a high correlation between drinking from the Broad Street pump and contracting cholera. This did not prove that drinking the water caused cholera; he demonstrated that the acts were correlated.

This type of research does not prove causation, but it is a powerful, scientifically valid tool to analyze data and come to an improved understanding of the world around us. Despite the fact that epidemiology studies can show only correlations and cannot prove causality, scientists, physicians, public health experts, and people in general tend to rely heavily on epidemiological studies for guidance on public health issues. Cancers, like other diseases including Alzheimer's, take a long time to develop—too long to run experiments in a laboratory. Scientists have used the tools of epidemiology to demonstrate, for instance, the increased correlation of smoking tobacco and the incidence of lung cancer. In fact, the government warnings against smoking were first issued based largely on epidemiological evidence, decades before the causal relationship between tobacco smoking and lung cancer was established.

When we approach the question of the health effects of EMF on human beings, we must similarly rely heavily on the tools of epidemiology and the examination of larger populations of individuals.

CELL PHONES AND CANCER

Epidemiological studies have begun to show us that low-frequency, non-ionizing electromagnetic radiation correlates with cancer. Exposure to even low-frequency, non-ionizing EMF (including radiation from cell phones and WiFi devices), as well as extremely low–frequency, non-ionizing electromagnetic radiation (such as comes from power lines), may be carcinogenic. We see this in a large number of epidemiological studies that demonstrate a correlation between specific types of exposure to EM radiation and the incidence of specific types of cancer in large populations.

In recent years, a significant share of the research into EMF and cancer has investigated the impact of cell phone radiation. In 2009, the *Journal of Clinical Oncology* published the findings of a team of seven scientists, who reviewed 23 epidemiological studies on the link between cell phone use and cancer. The team concluded that

> although as a whole the data varied, among the 10 higher quality studies, we found a harmful association between phone use and tumor risk. The lower quality studies, which failed to meet scientific best practices, were primarily industry funded.
>
> The 13 studies that investigated cell phone use for 10 or

> more years found a significant harmful association with tumor risk, especially for brain tumors, giving us ample reason for concern about long-term use.[5]

The higher-quality studies demonstrated a positive correlation between cell phone use and cancer, and the longer-term studies demonstrated an even stronger link. The point about length of exposure is a very important one. Cancers do not form overnight. In almost all cases, cancerous tumors take many years or decades to form and metastasize, and in many cases they result from *extended* exposure to carcinogenic agents. These results suggest that, as with tobacco smoke, cancer may be a long-term result based on repeated and/or extended exposures to EM radiation sources. It is for this reason that studies that conclude there is no link between cell phones or other sources of EMF and increased rates of cancer based on short-term exposure and evaluation are highly suspect.

On the contrary, there is a strong and increasing body of evidence that demonstrates the relationship between EMF exposure and cancerous outcomes. A 2007 review of 16 studies on this subject found that the studies all demonstrated increased risk of brain tumors known as glioma and acoustic neuroma (cancers that develop on the nerves associated with hearing and balance that run along the side of the face) among cell phone users.[6] Compiling the data across studies for 10 years and greater, the risk of developing an ipsilateral glioma tumor (a tumor on the same side of the head as the cell phone is used) was elevated 240%. The incidence of ipsilateral tumors reinforces the connection between radiation and cancer, as these tumors develop in the precise area where exposure to cell phone radiation has been most intense.

Note that the 240% increased risk is an average across studies, with some studies demonstrating much higher risks. In one of the cited studies, researchers concluded that there is an "increased risk of acoustic neuroma associated with mobile phone use of at least 10 years' duration," with a 90% increased risk for the auditory nerve cancer and an astounding 390% greater risk when restricting to ipsilateral use.[7]

This study also found that there was not an increased risk in those who used cell phones for fewer than 10 years. However, while this may be true for adults, other research indicates that children are much more sensitive to shorter exposures. In 2009, Dr. Lennart Hardell reported that children who begin using mobile phones at ages younger than 20 have a 520% elevated risk of developing glioma—even after just one year of use (this is compared to a 140% elevated risk across all ages).[8]

Other recent research out of Israel reinforces these findings. Israelis are heavy cell phone users who, on average, increased cell phone usage six-fold between 1997 and 2006.[9] As Dr. Siegal Sadetzki of Tel Aviv University explains, this population provides an excellent context in which to examine the potential for low-frequency EMF to cause cancer in human populations.[10] In 2008, Sadetzki and her colleagues published a study in the *American Journal of Epidemiology* that concluded that heavy cell phone users (those who use cell phones at least 22 hours a month) were at least 50% more likely to develop cancer of the parotid gland (one of the salivary glands) than those who never or rarely used mobile phones. Sadetzki's results, consistent with Hardell's findings, demonstrated a higher incidence among those with ipsilateral use, reinforcing the link between the use of cell phones and the occurrence of cancer.[11] Sadetzki concludes that "this unique population has given us an indication that cell phone use is associated with cancer."[12]

Given Sadetzki's findings on the link between cell phone radiation and parotid cancer and the popularity of cell phones in Israel, one would expect to see a notable increase in cancer of that type, corresponding to the rollout of this technology. And this is precisely what a review of the data demonstrates. A review of deaths between 1970 and 2006 from Israel's National Cancer Registry found that "the total number of parotid gland cancers in Israel increased 4-fold from 1970 to 2006 (from 16 to 64 cases per year), whereas other major salivary gland cancers remained stable."[13]

This result can be thought of as a control experiment. During this period, when parotid gland cancer was increasing in Israel along with cell phone use, cancer of the submandibular and sublingual glands (shielded from cell phone radiation by the jaw bone and tongue, respectively) remained constant. This is illustrated in the figure below, which plots the incidence of parotid (◆), submandibular (■), and sublingual (▲) gland cancers between 1970 and 2006.

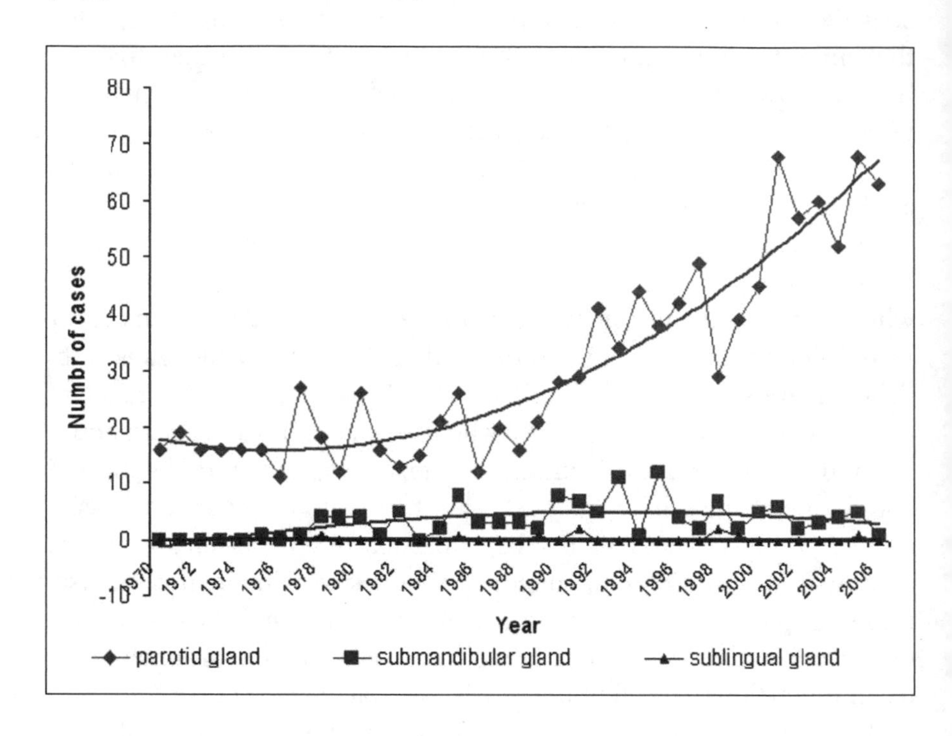

TOWERS

In studying the cancer and health risks of cell phones, it's important to remember that there is also radiation from the towers and antennas needed to transmit the signals. Whereas use of a cell phone is discretionary and individuals can opt to minimize or eliminate their exposure to radiation from mobile devices, these towers are always on and broadcasting, emitting RF and MW into the environment, and radiating everyone within range—whether or not they use a cell phone. This, of course, includes the many infants and young children who are not yet cell phone users. People have no choice as to whether or not to be exposed to this radiation. In a sense, tower radiation can be viewed as the cell phone equivalent of second-hand smoke.

Industry and related organizations acknowledge the presence of RF from these towers but they maintain that public exposure to the radio waves from cell phone tower antennas is slight.[14] For example, they point out that the antennas are mounted high above ground level with the result that the RF loses much of its intensity before it reaches people. Despite these claims, several recent studies indicate an increase in cancer associated with proximity to cell towers. And the closer one is to the tower, the greater the risk of developing cancer. In fact, some people live in apartment buildings with these transmission towers immediately above, on the roof, and as we can see from the example of England's Tower of Doom, the results can be tragic.

The Tower of Doom is a pair of masts, or towers, owned by the British firms Vodafone and Orange that were erected on top of an apartment building in 1994. In the following years, the local residents, and particularly the dwellers of the apartments in the buildings directly under the masts, began noticing an increased incidence of negative health effects, including cancer, which they attributed to the towers. Among the residents of the building who were affected are John Llewellin, who died of bowel cancer; Barbara Wood, Joyce Davies, and Hazel Frape, who all died of breast cancer; Barbara Watts and Phyllis Smith who both developed breast cancer; and Bernice Mitchell who developed uterine cancer. The cancer rate on the top floor of the apartment building was 10 times higher than the national average, with inhabitants of five of the eight apartments becoming ill. This was, in fact, a *cluster event* (several cases of cancer in the same location), with seven documented cases of cancer in a group of 110 residents. Eventually, after years of public pressure and action, Orange agreed to remove their tower. Vodafone, in contrast, has not, and the South Gloucestershire Council remains unable to force removal because current safety standards do not indicate that cell phone tower radiation is dangerous.[15]

A 2011 review of the research published on the question of long-term exposure to low-intensity MW radiation (such as that to which you may be exposed by nearby towers) noted demonstrable carcinogenic effects, which typically manifested after extended exposure of 10 years or more. Unfortunately, these low levels of MW radiation are well within the

ICNIRP (the International Commission for Non-Ionizing Radiation Protection, of the World Health Organization) safety standards (discussed in chapter 10).[16] On the specific question of radiation from cell towers, this review noted that even one year of exposure led to a "dramatic" increase in cancers in nearby residents. One study compared cancer cases among people living up to 400 meters (less than a quarter of a mile) from a base transmitting station and people living farther away. After ten years, the group close to the base station had over three times the rate of cancer relative to those living at a greater distance.[17]

Similar findings have been reported in studies in Brazil, where the greatest accumulated cancer incidence was among those exposed to power densities as high as 40.78 $\mu W/cm^2$. The reported incidence was 5.83 per 1,000. Those farther away who were exposed to a power density of 0.04 $\mu W/cm^2$ (levels 1,000 times lower than the group with the highest exposure) had a lower cancer incidence of 2.05 per 1,000.[18] These studies, and others like them, indicate that towers are a significant component in the risks associated with cell phones.

NOT JUST CELL PHONES

So far, we've discussed the possible health risks of cell phones and their associated technology. But cell phones are only one of many devices that produce and transmit radio-frequency EMF. The modern world is pervaded by "EMF producers"—microwave ovens, computers, radio and television broadcasting, to name just a few—and they have a wide range of characteristics.

Dr. Neil Cherry was an environmental scientist from New Zealand who spent a good deal of time researching the questions of the health effects of electromagnetic radiation. In one study, Dr. Cherry investigated the health risks associated with television and FM radio broadcasting antennas—which broadcast EMF at a lower frequency than cell phones and cell phone towers. As part of his study, Cherry examined the incidence of childhood cancers over half a century among those who lived close to the Sutro Tower broadcasting antennas in San Francisco between the years 1937 and 1988. By plotting the occurrence of 123 cases of cancer among 50,686 children, he demonstrated that living closer to the tower correlated with a higher incidence of childhood cancer and that the risk for such cancer dropped significantly with increased distance from the antennas. Overall, the incidence of childhood cancer was quite high—especially considering that the measured EMF at three kilometers (where the relative risk was about six) was approximately *a thousand times lower* than the currently accepted safe level.[19] Yes, you read that correctly! The relative risk is high (~6) even at a power density that is approximately *a thousand times lower* than the currently accepted safe level.

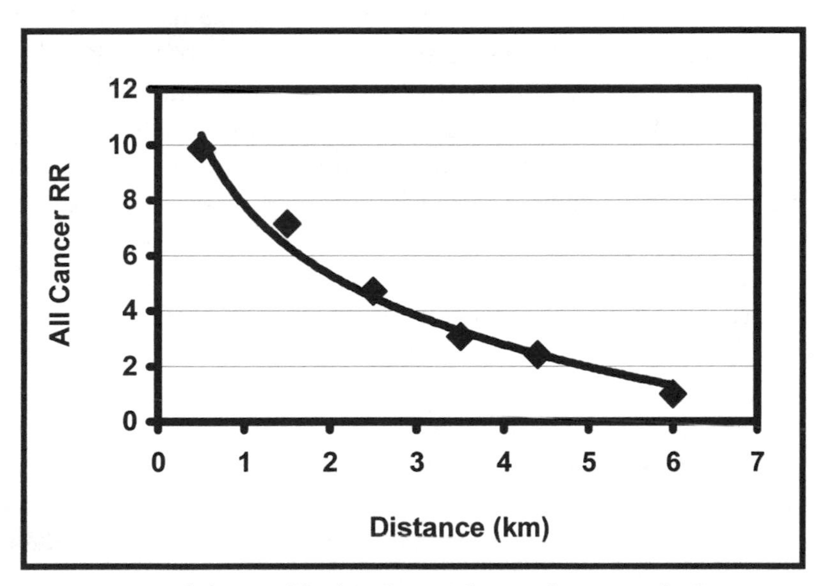

Sutro Tower data, 1937–88. The relative risk (RR) of cancer vs. the distance in kilometers from the antenna.
Neil Cherry PhD, "Childhood Cancer in the Vicinity of the Sutro Tower, San Francisco," Human Sciences Department, Lincoln University, September 19, 2002.

Dr. Orjan Hallberg of the Karolinska Institute in Sweden published a series of studies between 2002 and 2008 that examined the carcinogenicity of FM transmitters. Hallberg noted that rates of melanoma in Sweden and other Nordic countries had been on the rise since 1960—in contrast to the previous 50 years in which rates of melanoma incidence had been stable. Hallberg hypothesized that FM transmitters (which had been introduced in the Nordic countries in the 1950s) might be involved. His studies first demonstrated that rates of melanoma increased with the length of exposure to FM-frequency EMF radiation, from which he concluded that "melanoma is associated with exposure to FM broadcasting."[20] He pinpointed 1955 as the point in time when these stresses were introduced into the environment[21] and developed a model to explain the increasing incidence.[22] He subsequently released a model explaining that "reduced efficiency of the cell-repairing mechanisms [such as those we discussed in the previous chapter] is capable of explaining the increasing trends of melanoma incidence that we have been noticing since the mid-20th century."[23]

Adding evidence to support his conclusions, Hallberg noted that whereas in prior generations melanomas were generally limited to those areas exposed to the sun, increasingly such cancers were being found all over the body (as would be expected if these cancers were caused by exposure to EMR, rather than solar radiation).[24]

As with other studies we've seen, Hallberg demonstrated a link between the degree of EMF exposure and the incidence of cancer, finding that "those who had four FM-radio or TV towers covering their residential area are more than twice as likely [to develop melanoma] as those who had one." This is what is known as a *dose-response* relationship, in which an increased dose yields an increased response. Data that demonstrates this dose-response relationship is generally considered more reliable, as a stronger exposure leads to a stronger effect, increasing the correlation between exposure and effect that the data demonstrates.

POWER LINES

Communication towers are not the only technical infrastructure on which television, FM radio, and cell phones rely. These and other electronic devices are similarly dependent on access to electrical power and that access comes via power lines. The lines transmit the power generated at the main power stations and bring it to substations for distribution in neighborhoods.

In a few locations, these power lines are placed underground, thus reducing exposures at ground level. Because underground installation of power lines is more expensive, only a few urban centers such as New York City and San Diego have buried lines. One 2005 study concluded, for example, that the cost of converting aboveground power lines to underground in Virginia would total $800,000 per mile, or $83.3 billion.[25] More typically, power lines run above ground, positioned to affect much wider areas.

The voltage (the force driving the electric current) is often so strong that you can hear a crackling noise coming from the big cylindrical transformers. Electric power transmission occurs at high voltage and is accompanied by strong magnetic and electric fields. Here, once again, the epidemiological evidence is that exposure to these sources of low and extremely-low-frequency non-ionizing EMF is associated with increased rates of cancer.

All of this power comes from the power grid. While this grid is ubiquitous in the United States today, it was rolled out in different parts of the country at different times—generally, urban areas preceded rural regions and the northern United States preceded the South. By examining official archives and death records across these different regions from different periods of time, Dr. Sam Milham was able to show how increases in the death rate correlated with the onset of electrification, independent of where it occurred in the country. In one case, building upon existing, accepted research that childhood leukemia showed a peak incidence in children three to four years old,[26] Milham and his colleague Eric Ossiander published a study showing that the appearance of this characteristic peak correlated with the introduction of electrification. In other words, the trend of childhood leukemia developing at a certain age is a modern one, which followed the introduction of the power grid, and this peak does not exist in unelectrified regions, such as sub-Saharan Africa.

While the entire power grid generates ELF radiation, high voltage power lines generate much stronger ELF fields than the power lines that run to your home. The EMF levels radiated by such lines are particularly dangerous. This link between EMF radiation from power lines and childhood leukemia was pointed out in 1979 by Dr. Nancy Wertheimer and Ed Leeper, who studied the possible carcinogenic effects of exposure to EMF from power lines in certain homes in Colorado. They concluded that growing up in a home surrounded by such high voltage electrical cables was, in fact, associated with an increased incidence of childhood leukemia. "The finding was strongest for children who had spent their entire lives at the same address, and it appeared to be dose-related." They did not reach a conclusion as to why this correlation exists, but did mention "AC magnetic fields" as one potential reason.[27]

These findings and those of many other similar studies led the World Health Organization (WHO) in 2002 to include ELF (power-frequency EMF) among the possible causes of childhood leukemia.[28] (WHO gave a similar evaluation regarding radio-frequency EMF and cancer in 2011,[29] again relying heavily on information from epidemiology studies. These two decisions, warning of the possible health effects of EMF, cover almost all frequencies in the non-ionizing range of electromagnetic radiation.)

Some scientists previously believed EM radiation could only damage humans if the radiation was sufficiently intense to cause heating of the tissues. This theory, which is often referred to as the *thermal criterion*, has now been roundly discredited by many studies in which biological effects have been observed at intensities far too low to cause any measurable heating effect.

While cancer is a serious condition, it is just one set of human diseases that are, in many cases, caused by environmental stresses. If a force in the environment is strong enough to potentially cause cancer, then that force is also powerful enough to cause many other types of damage. In the next chapter, we'll examine some of the other known health effects of electromagnetic radiation.

Chapter 6

OTHER HEALTH EFFECTS OF EMF

The traditional Amish Mennonite Christians live a life that many of us would consider austere, with strict rules that govern everything from how they travel to what clothes they wear. The Amish first immigrated to the United States in the early 18th century from areas in and around Germany, and settled in Pennsylvania (though they now reside in many other parts of the United States, including Ohio and Indiana). They have maintained their way of life for 300 years. The Amish value family and interactions with friends, and they shun many modern conveniences like telephones and cars.

Amish in typical dress, horse and buggy.

In fact, among those who belong to the Old Order Amish, all use of electricity is prohibited. And as we examine the public health impact of the increasing doses of EMF to which most of us are exposed on a daily basis, it's useful to have such a group to which we can compare ourselves. They are obviously not a control group, but the many ways in which their lives differ from the general population provide interesting insights.

For example, when we look at cancer rates, the Amish have notably lower levels than the rest of the American population. Dr. Judith Westman studied the death rates from cancer in the Ohio Amish, expecting to find a higher incidence of cancer stemming from intermarriage and other social factors. Instead, upon detailed review of the records of the Ohio Cancer Registry and household interviews, Westman calculated that the Ohio Amish had a stunning 40% lower incidence rate for all cancers when compared to the average Ohioan. She published her findings in 2009, noting that other known aspects of the Amish lifestyle (such as the prohibition of tobacco) could not explain these results.[1]

Other health differences are also observable. According to Dr. Sam Milham (whose work on the effects of electrification was referenced in the previous chapter), the Old Order Amish have lower rates of heart disease, diabetes, and suicide than the population at large. These Amish also have very low levels of neurodegenerative disease, and there is not one single documented case of ADHD (Attention Deficit Hyperactivity Disorder) among the Old Order Amish![2]

DISEASES OF CIVILIZATION

In broad terms, the Old Order Amish—who are not exposed to the same types and levels of man-made EMF pollution—do not suffer from what Milham refers to as the "diseases of civilization," which include Alzheimer's, infertility, depression, and heart disease. Cancer is certainly a dangerous disease and an on-going public-health concern, and it justly receives a lot of attention. But it's not the only set of diseases whose incidence has either appeared, or notably increased, along with the introduction of more and more sources of EMF radiation in the environment.

The Amish are not the only group we can look to for such a

comparison. As I explained in the previous chapter, the process of electrification proceeded at different rates in different parts of the United States, and as Dr. Milham reviewed historical death records from these different regions, he was able to note differences in the incidence of multiple diseases of civilization, including cancer, cardiovascular disease, diabetes, and suicide—all of which tended to increase with the onset of electrification. There is an increasing body of evidence corroborating Milham's conclusions.

ALZHEIMER'S

Alzheimer's disease is a neurodegenerative condition that destroys the neurons in one's brain—and for which there is currently no known cure. This debilitating disease remains poorly understood, despite decades of research. But there is a growing body of scientific evidence indicating that exposure to low-frequency EMF (such as that emitted by power lines) significantly increases the risk of developing Alzheimer's disease.

Using mortality and census data between the years 2000 and 2005, a group of Swiss researchers investigated the relationship between exposure to 220–380 kV high voltage power lines (which run near homes and emit extremely-low-frequency EMF) and neurodegenerative conditions such as Alzheimer's. They concluded that persons living within 50 meters (less than one city block) of such lines had a 24% higher risk of developing Alzheimer's disease than persons living more than 600 meters away (about a third of a mile).[3] A dose-response relationship was also noted, with the risk of Alzheimer's increasing steadily with each year of exposure. Those who had lived near the lines for at least five years were over 50% more likely than average to develop Alzheimer's, and those who had lived near these high voltage power lines for 10 years had twice the normal risk.

High voltage power lines are one known source of high-powered ELF. Industrial machinery commonly used in many professions is another. A 1995 study led by Dr. Eugene Sobel at the University of Southern California School of Medicine found that those in certain careers who are routinely exposed to a significant amount of extremely-low-frequency EMF were at a notably elevated risk of developing Alzheimer's. Individuals in specific careers (mainly seamstress, dressmaker, and tailor), frequently exposed to EMF levels over 50 mG from sewing machines, had three times the risk of developing Alzheimer's (3.8 times higher in the case of women).[4]

OTHER NEURODEGENERATIVE DISEASES

Alzheimer's disease is not the only neurodegenerative disease that Milham has noted. Another is amyotrophic lateral sclerosis (ALS), commonly known as Lou Gehrig's disease. ALS is less common than Alzheimer's, affecting an estimated 5.4 million people worldwide[5] and causing sufferers to gradually lose control of the voluntary muscle movements of their bodies due to the deterioration of nerve cells in the brain and spine.

A pair of 1998 studies led by Dr. David Savitz from the University of North Carolina demonstrated career-specific risk of neurodegenerative diseases resulting from increased EMF exposure. In one, Savitz calculated that those in electrical occupations were two to five times as likely (depending on the specific occupation) to develop ALS and noted that power-plant operators, in particular, were at significantly elevated risk (up to five times greater than normal) of developing ALS as well as Alzheimer's and Parkinson's (another in the class of neurodegenerative diseases).[6] In the second study, Savitz found that electric utility workers were at twice the risk of developing ALS as the general population.[7]

Other studies arrived at similar findings. Danish electrical workers have been found to be at twice the risk of developing ALS.[8] Based on a review of health records of 718,221 people, researchers concluded that welders in Sweden (who are exposed to extremely high levels of ELF from welding equipment) were at four times the risk of developing Alzheimer's and over twice as likely to develop ALS when compared to nationwide averages.[9] Another study of dementia in 931 elderly Swedes also demonstrated that men in careers with high levels of EMF exposure are at over twice the risk of developing Alzheimer's and dementia.[10]

While certain occupations have clear risks, there are some careers that you might be surprised to find have high levels of EMF exposure. One such career is railway work. A Swiss study included 20,141 individuals who had been employees of the Swiss railway in the years from 1972 to 2002 and concluded that the train drivers (the employees with the highest level of exposure to ELF) were at more than three times the normal risk of developing Alzheimer's. Röösli also demonstrated that, with each year of exposure, the risk of developing Alzheimer's or ALS demonstrably increased (indicative of a dose-response relationship).[11]

Results such as these, noting career-specific risks from EMF exposure in the workplace, have been reported in the United States as well. Using death-certificate data from 22 states in the period from 1992 to 1998, a team working at the National Institute for Occupational Safety and Health in Ohio found indications of elevated risk of Alzheimer's stemming from occupation.[12] These researchers identified bank tellers, clergy, aircraft mechanics, and hairdressers as the occupations with the most elevated risk of Alzheimer's. In particular, they noted that individuals who spent an extended period of time exposed to 60 Hz EMF (the EMF frequency emitted by power lines in the United States, and much of the world) were 87% more likely to develop Alzheimer's or Parkinson's.

Recognizing the increasing body of research demonstrating a connection between ELF exposure and Alzheimer's, in 2007, Dr. Ana Garcia of the University of Valencia in Spain released an evaluation of 14 different studies in this area. While Garcia found some variability in the studies, she concluded that the "available epidemiological evidence suggests an association between occupational exposure to ELF-EMF and AD [Alzheimer's disease]."[13]

There is a clear and continually growing body of evidence that exposure to extremely-low-frequency EMF radiation from power lines, electrical fixtures, and other appliances—in the workplace and at home—increases the risk of developing Alzheimer's and other neurodegenerative diseases. Because of their often debilitating nature, these diseases are of great concern to health workers and the public at large. And the elevated risk of neurodegenerative diseases from EMF exposure is among the more alarming of the noncancerous impacts of EMF. Still, it is only one of many afflictions that epidemiological studies have associated with EMF exposure.

MALE INFERTILITY

Another significant health issue affecting a much larger segment of the population is male infertility, which has seen a dramatic rise. Data indicate that sperm counts have decreased by half in the past 50 years and are continuing to fall.[14] Currently 7% of men worldwide are infertile and 45% of men are subfertile.[15] While the biggest EMF source of elevated risk for Alzheimer's appears to be ELF associated with power lines, it appears that mobile phones are the key source of EMF that is linked to increased rates of male sterility.

A 2011 review of the various scientific studies published on EMF and male fertility concluded that men who used cell phones experienced "decreased sperm concentration, motility, abnormal morphology, and viability" and that sperm exposed to RF radiation in a laboratory have "decreased motility, morphometric abnormalities, and increased oxidative stress."[16]Decreased motility reflects a reduced ability of sperm to move toward eggs and is considered an indicator of sperm quality. Morphometric abnormalities mean that the sperm are of an abnormal shape and form. And an increase in oxidative stress reflects a reduced ability of cells to repair damage and is linked with many diseases including cancer. In short, RF radiation significantly damages sperm. The review also found that the longer one used a cell phone, the higher the risk of these abnormalities.

Among the work covered in that 2011 study was some startling research conducted by Dr. Ashok Agarwal and his colleagues at the Cleveland Clinic, who are scientists studying issues of infertility, including the effects of cell phone use on men's sperm and semen quality. They divided over three hundred infertile men into four groups. One group did not use a cell phone, the second used it for two hours a day or less, the third for two to four hours a day, and the fourth group for more than four hours a day. Those in the highest-usage group had 40%

lower sperm counts than those who did not use cell phones during the study. The researchers also found dose-response relationships between cell phone radiation exposure and sperm count, motility, viability, and morphology.[17]

Agarwal's studies on EMF and sterility are not restricted to epidemiology—he has been able to approach this subject in a laboratory setting as well. In one study the researchers separated semen samples into two groups: those that received radiation equivalent to speaking on a cell phone for one hour and those that were not exposed to such radiation. The semen exposed to the radiation showed "a significant decrease in sperm motility and viability."[18]

Such findings are borne out in studies from other researchers. A group of scientists in Australia dramatically reinforced the relationship between cell phone radiation exposure (measured in SAR, specific absorption rate, discussed in greater length in chapters 3 and 10) and sperm health, noting that cell phone radiation decreases "the motility and vitality of these cells." The published data showed a strong inverse correlation between SAR and sperm vitality, clearly demonstrating a dose-response relationship.[19]

BLOOD-BRAIN BARRIER

While many of the studies I am citing are from the last twenty years, the health effects of RF and microwave exposure have been examined for many decades. In one of the earlier studies published in 1975, Dr. Allan Frey, a neuroscientist working for General Electric, examined changes that occurred with exposure to microwaves, focusing on structures that control the composition of the fluid that surrounds and protects the brain, the so-called *blood-brain barrier* (BBB).[20]

The brain is encased in a bone called the cranium. Between the bone and the brain, there is a liquid—the cerebrospinal fluid (CSF)—that protects the brain, cushioning it from any impact. In the absence of the CSF, any blow to the head could be transmitted across the bone directly to the brain. The fluid works to lessen the impact, thereby shielding the brain from physical injury. Cells called *endothelial cells* line the brain's blood vessels (capillaries), allowing certain substances to pass from the blood into the CSF. These cells form the blood-brain barrier (BBB) by separating the blood circulating in the brain from the CSF compartment. The BBB prevents many otherwise-harmful substances in your bloodstream (such as viruses and bacteria) from entering your CSF while permitting access for nutrients and other essential components that your brain needs to function.

Frey was interested in whether EMF exposure would impair the BBB. He studied two sets of rats: One set, which he exposed to 1.9 GHz radiation (comparable to a cell phone) for two hours and a second set, which did not receive that radiation exposure. First, he injected a fluorescent dye into the circulatory system of each unexposed rat. As expected, the dye spread very quickly into all the tissues—except the brain. Then he injected the same dye into the exposed rats, and within a matter of minutes—from this single exposure—the dye had leached into the brain. This meant that the BBB had leaked and that large molecules such as proteins and hormones that were previously excluded now crossed into the fluid bathing the brain. It is important to note that Frey's research revealed this damage with very short-term exposure, suggesting that abnormal changes begin almost immediately. Although Frey was unable to continue this research (for reasons discussed in detail in chapter 8), his results have been confirmed by others such as Professor Leif Salford in Sweden.[21]

It has since been shown that some brain cells die shortly after such leaks occur.[22] Even if neuron death does not occur, the breach of the BBB is very dangerous. The BBB, for example, helps prevent viruses and bacteria from entering the brain. The environment of the brain must be highly stable if nerve cells are to function properly. Breaching of the blood-brain barrier interferes with this stability.

MELATONIN PRODUCTION

Another well-established, but invisible, effect of EMF exposure that creates increased risk for disease is disruption of your body's melatonin production. Melatonin is a hormone produced from serotonin by the pineal gland (located in the brain) that helps regulate our sleep patterns through control of the circadian cycle. Melatonin is also associated with your ability to learn, to fend off damage from free radicals and other forms of aging, and other key immune-system functions including your body's ability to defend against cancer. In addition, melatonin levels in your body correlate with serotonin levels, which are related to human diet, metabolism, and even depression. Worryingly, by 2000, there were already 15 different studies demonstrating that ELF, RF, and MW radiation suppresses your body's ability to produce melatonin.[23]

The first such study was conducted in 1989 by S. G. Wang, who demonstrated that humans exposed to RF and MW fields displayed increased serotonin levels in a dose-response relationship, which indicated decreased melatonin production.[24] The following year, Dr. Barry Wilson led a team at Pacific Northwest Laboratory and the University of Montana that examined a similar question, this time using ELF instead of RF/MW. Wilson recruited a group of 28 volunteers who slept with electric blankets at night for eight weeks and analyzed the melatonin levels in their nighttime urine. From the results, which included "significant changes" in the melatonin levels in a quarter of these volunteers, Wilson hypothesized that exposure to ELF "can affect pineal gland function in certain individuals."[25]

Dr. James Burch led a team investigating the melatonin levels in electric utility workers exposed to generally high levels of 60 Hz ELF. Burch measured the exposure of these workers over a three-day period (capturing their exposure at work and at home) and found reduced melatonin levels, noting the greatest reductions when "exposures both at work and at home were combined."[26]

While electric blankets are a high source of ELF and industrial ELF exposures in the workers studied by Burch are even higher, Dr. Scott Davis of the Fred Hutchinson Cancer Research Center in Seattle was more interested in very low levels of ELF, which he believed was more in line with levels of ELF exposure in the general population. His research, published in 1997, demonstrates that low-level ELF exposure leads to decreased nocturnal melatonin production in a dose-response relationship. These subjects demonstrated significant dose-response results, with a two-fold increase in ELF exposure reducing melatonin levels 8%; tripling the ELF exposure reduced nocturnal melatonin levels by 12%; and a four-fold increase in ELF exposure led to a 15% reduction.[27]

While much of the early research into EMF and nocturnal melatonin production centered on ELF, more recent research has also focused on the impact of RF and MW frequency ranges of EMF. Burch explored this relationship in a 2002 paper that studied utility workers—this time focusing on the impact of the use of cellular telephones on their bodies' ability to produce melatonin at night. He measured EMF exposure of these workers over a three-day period and adjusted the results to account for ELF exposure, so that he could focus on the impact of cell phone EMF (alone, and in combination with ELF). From the results, Burch concluded that daily use of cell phones for more than 25 minutes led to a significant drop in nocturnal melatonin production. What's more, Burch noted that exposure to ELF potentiated this effect. In other words, exposure to 60 Hz EMF at home or work increased the likelihood of reduced melatonin production from exposure to cell phone radiation. The effects of these exposures across the EM spectrum accumulate and interact.[28]

DEPRESSION AND SUICIDE

As melatonin is involved in so many important bodily functions, the health effects of reduced nocturnal melatonin production resulting from exposure to low-frequency, non-ionizing forms of EMF radiation, can manifest in different ways. Many believe this melatonin-suppression effect is part of the underlying mechanism by which EMF increases risk of depression—and even suicide.

In an 18-month period from 2007 to 2008, 22 minors killed themselves in Bridgend, South Wales, giving the town the moniker "Britain's suicide capital." Upon further investigation, Dr. Roger Coghill discovered that all 22 lived closer than average to a cell tower. While the average Briton lives 800 meters, or about 2,625 feet, from a tower, these suicide victims lived, on average, only 356 meters (approximately

1,170 feet) from a tower—less than half the distance of the average Briton. As Coghill said in an interview with Britain's *Daily Express*, scientific research points "to the fact that exposure to mobile radiation can lead to depression," and he believes that circumstantial evidence exists that towers are responsible for causing depression in all 22 who died from suicide.[29] There is a large body of evidence that Coghill can cite to support his claims.

In one study, researchers estimated cumulative ELF exposures over a 20-year period for 12,063 Finnish individuals. Each of these subjects then responded to a 21-item Beck Depression Inventory (an accepted metric used to assess depression). The results demonstrated that the risk of severe depression was increased 470% for people who lived within 100 m (about 330 feet) of a high-voltage power line.[30]

While that study covered high-voltage power lines, similar conclusions were reached the year before by a British team that studied residents of Wolverhampton, England, finding a significantly higher risk of depression for those individuals residing closer to residential power lines with levels of ELF radiation stronger than 50 Hz outside of their homes.[31] Another team of researchers in the United States, led by Dr. Charles Poole, found that symptoms associated with depression were 2.8 times more prevalent for those who live close to the power lines.[32]

Depression, in severe cases, is among the potential causes of suicide. If exposure to EMF is shown to cause depression, we might also expect to see a link between exposure to electromagnetic radiation and incidence of suicide. Sadly, this is what many studies reveal. Dr. Maria Reichmanis led a team that demonstrated in 1980 a significant correlation between suicides and magnetic field strength. The researchers investigated 590 of the 651 suicides reported by coroners in England's West Midlands region between 1969 and 1976. The team measured EMF levels at each of the home addresses of the suicide victims and also at 594 control addresses for comparison. They found that the magnetic field at the homes of those who died from suicide was higher than at the control addresses. Particularly, the proportion of suicides found in the "high" and "very high" classes of EM exposure were 40% greater than in the equivalent control subjects.[33]

A group of US researchers published findings from their research on a similar question, this time focusing on workplace, rather than residential, ELF exposure in a study of 138,905 male electrical workers. The results demonstrated that power-line workers were at a 59% greater risk of dying from suicide. Electricians had more than double the risk. Even stronger associations were found when looking only at those younger than 50 years of age. These individuals were at over a three times increased risk of death by suicide than national averages.[34]

Depression and suicide are extreme examples of emotional disruption. Science also indicates that EMF can trigger less damaging, but still disruptive, cognitive dysfunction. Certain types of EMF exposure can impair the way in which we think. A series of three papers was published in *Bioelectromagnetics* between 2006 and 2011 by Drs. Roy Luria, Ilan Eliyahu, and Ronen Hareuveny following their studies of the cognitive effects of cell phones. In each, right-handed subjects were split into two separate groups: one group was exposed to cell phone radiation on the left side of their heads, and the other group received exposure on the right side of their heads. The subjects were then asked to perform a series of tasks that rely on cognitive function and memory, requiring either a left-handed or right-handed response. The results indicated "that the exposure of the left side of the brain slows down the left-hand reaction time" and that the same is true for right-hand reactions after exposure on the right side of the brain.[35]

EYES AND EARS

Some of the earliest research into the health effects of non-ionizing EMF exposure was that of Dr. Milton Zaret, who passed away in 2012 at the age of 91. In the 1950s, Zaret (an ophthalmologist by training) explored the risks to the eye from microwave radiation (with which the public was just starting to come into contact, from radar, microwave ovens, and diathermy machines used in physical therapy). He theorized the existence of *microwave cataracts*—a clouding of the eye caused by exposure to microwave radiation. Unlike normal cataracts, which form toward the front of the eye, Zaret explained that microwave cataracts form on the posterior capsule at the rear of the eye (because microwave radiation can reach through the eye). These cataracts could develop from chronic low levels of MW exposure, not just from strong exposures. (Unfortunately, due to military disinterest in studying negative health effects resulting from exposure to EMF from military equipment, Zaret lost all of his military contracts, and no subsequent attempts have been made to re-create his results.)[36]

Another of the early researchers into the nonthermal effects of EMF exposure was Dr. Allan Frey—the same scientist who performed the blood-brain barrier studies described earlier in this chapter. In the 1960s, Frey discovered an effect of microwave radiation on human hearing that became known as the *Frey Effect* (also referred to as *microwave hearing* or the *microwave auditory effect*). Frey exposed a series of subjects to extremely low frequencies of pulsed EM radiation from a transmission antenna. The subjects were about 300 hundred feet away from the antenna, and by turning on the transmission, Frey was able to induce the perception of sound in these subjects. Many of these subjects reported hearing a buzzing or clicking noise immediately when the antenna was activated. The subjects reported additional effects including headaches and dizziness.[37] Frey's results have been replicated many times and later studies demonstrated that this auditory-system response occurs when exposed to EM radiation in the frequencies of 200 MHz to 3 GHz.[38]

These effects may seem minor when compared to diseases discussed in this chapter, such as Alzheimer's and ALS. However, they are indicative of an even broader range of biological responses to electromagnetic radiation. EMF exposure is not just linked to cancer and Alzheimer's. There is good reason to believe that EMF exposure is associated with a wide number of negative health outcomes.

CONCLUSION

As we learn about the risks of EMF exposure and its wide-ranging health effects, which are scientifically demonstrated in humans, we begin to grasp the seriousness and enormity of the problem. As Martin

Halper, the EPA's Director of Analysis and Support, says, "I have never seen a set of epidemiological studies that remotely approached the weight of evidence that we're seeing with EMFs. Clearly there is something here."[39]

Equally evident from the increasing mass of scientific evidence of the health effects of exposure to low frequency, non-ionizing electromagnetic radiation is that individuals respond differently to different doses of exposure and to different frequencies of EMF. For instance, low levels of ELF exposure may cause melatonin suppression in some individuals, but not in others. In others, the same level of ELF exposure may be linked to the occurrence of leukemia. We do not know or understand enough at the present time to predict outcomes. We can only demonstrate that these negative health effects are strongly linked (either by correlation in epidemiological research or by causation in laboratory experiments) to various types of EMF exposures widely considered to be safe.

It is clear that the health risks from man-made EMF are real, with a wide-ranging impact on people. This is the consensus of a growing number of responsible scientists and professional health-care workers. Of course, any environmental pollutant that is strong enough to cause this degree of damage to humans is bound to damage other animals and plants. As we'll see in the next chapter, this is what the science demonstrates.

Chapter 7

THE NONHUMAN IMPACT OF EMF

Skrunda-1 is a small ghost town located in eastern Latvia, a former Soviet Baltic state east of Russia. The 110-acre area, abandoned since 1988,[1] is located just to the north of Skrunda, a small village of just under 4,000 people.[2] Still standing today are 70 or so abandoned structures (schools, hospitals, barracks, and other similar types of buildings) that were once used by the Soviet military as part of their operation to detect US and NATO actions in Western Europe. Skrunda-1 was among the more than 40 hidden municipalities in the USSR that did not appear on any official maps and were referred to only by code names.[3]

While many relic structures remain, gone are the powerful Cold War radio and radar installations around which this secret military town was built. The first radar stations went up in 1964. A more powerful tower, known as a Dnepr, was erected in 1971, enabling the military to observe objects over 3,700 miles away—and over 1,800 miles into space. Construction on a third, even more powerful tower known as a Daryal began in 1984 and was to become the most important station for Soviet ability to monitor space.[4]

Due to the shifting winds of politics, though, the Daryal station was never completed. In 1991 Latvia had gained nominal independence from Russia, and in 1994, the Latvians and Russians agreed to a Russian withdrawal from Skrunda-1 by 1998. In this period, the still-incomplete Daryal tower was demolished, and in 1998, the Russians removed what they could from the secret military town, creating the ghost town.

By then, however, incalculable damage had already been done.

Constructing and operating Skrunda-1 may have provided a valuable strategic asset to the Soviet military, but it came at the cost of exposing Skrunda-1, the surrounding area, its inhabitants, and its nature to immense levels of RF radiation—with intensities up to 50 times higher than levels indicated as safe by ICNIRP and WHO[5]—for 34 years. In the years since the abandonment of Skrunda-1, numerous studies have been conducted documenting many of these effects on the surrounding environment.

Duckweed—tiny oval-shaped plants that float on the surface of slow-moving freshwater—has been found to live shorter lives and yield fewer offspring than normal.[6] The pine trees in the most exposed zones have fewer needles[7] and die at a younger age.[8]

And it is not just the plants!

Cows exposed to radiation from the Skrunda-1 towers have been found to have increased rates of DNA damage in the form of higher amounts of micronuclei in cells,[9] and significantly fewer pied flycatcher birds have nested in the RF-polluted region.[10]

Skrunda-1 may present an extremely focused example of the types of damage that low-frequency non-ionizing radiation may cause to the nature around us, but a growing body of research indicates that the increasing scale of electromagnetic pollution commonly found around us is having unforeseen consequences throughout the animal and plant worlds.

THE BIRDS AND THE BEES

Among the fauna suffering negative outcomes from EM exposure, we find both migratory birds and honey bees. One of the key ways in which RF and MW radiation is seen to affect these creatures is through disrupting their navigational abilities. They are among a set of animals that navigate using *magnetoreception* (or *magnetoception*), the ability to sense magnetic fields. Honey bees,[11] migratory birds,[12] salmon,[13] bats,[14] fruit flies,[15] even bacteria,[16] and some studies suggest, humans[17] are among those species that are shown to have a sense of the earth's magnetic field, which is used as a navigational aid. Magnetoreception is so strong in some birds that they are said to "see" the magnetic field, in the way that people see color.[18]

Magnetoreception is still poorly understood. There are two primary models that scientists use to explain the phenomenon. One involves *cryptochromes,* a set of proteins in many animals and plants known to be involved with circadian rhythms and found in the retinas of many birds.[19] Dr. Dominik Heyers at the University of Oldenburg in Germany and fellow researchers demonstrated that retinal cryptochromes in nocturnally migratory songbirds show high levels of communication with the brain during periods when these birds rely on magnetoreception to orient themselves.[20] Similarly, Robert J. Gegear at the University of Massachusetts Medical School and his fellow researchers demonstrated that cryptochromes are a critical component for magnetoreception in fruit flies.[21]

The other model involves *magnetite,* which is an iron-based mineral found in nature, sensitive to magnetic fields, including the earth's magnetic field. As with cryptochomes, magnetite has been found in many species of animals including some bees, birds, insects, fish, bacteria, and humans; and several studies indicate that magnetite is involved in magnetoreception. In 2001, Dr. Joseph Kirschvink reviewed the research on magnetite and magnetoreception and concluded that the ability to sense magnetic fields depends on an internal sensory system built from magnetite crystals inside the body.[22] Years earlier, Kirschvink had shown that there are large quantities of magnetite in honey bees and pigeons, and theorized that "when integrated by the nervous system," magnetite is "capable of accounting for even the most extreme magnetic field sensitivities reported."[23] Other research demonstrated that magnetite in mammals is sensitive to 60 Hz fields that are approximately *only 1/50th* the strength of the earth's magnetic field.[24]

While magnetoreception science is still in its infancy, the presence and importance of this sense to several animal creatures is well established. Researchers have identified two biological mechanisms (cryptochrome- and magnetite-based) that are likely involved with magnetoreception. Based on this, scientists are building an understanding of just how low-frequency EMF is impairing the ability of birds and bees to navigate their worlds, leading to reduced populations of both.

MIGRATORY BIRDS

Low-frequency EMF has been demonstrated to impair both cryptochrome- and magnetite-based systems in birds. A group of researchers led by Dr. Wolfgang Wiltschko at the University of Frankfurt demonstrated that birds use magnetite receptors when forming "maps" of their worlds. Studying the migratory Australian silvereye, the researchers exposed the birds to a pulse of magnetic field and noted that the exposed birds were thrown off course by the same type of EMF they rely on for this critical navigational information.[25] This built on work that Wiltschko and his coauthors (including Dr. Ursula Munro of the University of Indiana) released years earlier, demonstrating the presence and role of these "magnetite-based 'navigational maps'" in birds.[26]

Similarly, when exposed to magnetic fields in the range of 100 Hz to 10,000 Hz, robins demonstrated spatial disorientation. This effect was seen with field strengths of only 1/500th of the earth's magnetic field.

Whatever the actual systems are in birds that are impacted by EMF exposure (and certainly, scientists will continue to research this subject), some of the results of avian exposure to this type of disorienting radiation are becoming increasingly clear.

NAVIGATION

The most visible impact of avian disorientation is the increasing phenomenon of tower collisions—when birds fly into EM transmission towers (including electricity relay, cell, radio, and television towers) and die. The first reported avian fatality from a radio-tower collision was reported in Baltimore, Maryland, in 1949,[27] although eagle mortality from collisions with power lines were reported in 1922.[28]

The first long-term study in this area began in 1955 in northern Florida. By 1980, 42,384 avian deaths across 189 species were recorded[29]—65% of these deaths occurred in the autumn months (when these migratory birds were flying south), and 20% in the spring (when these birds reversed course, flying north).[30] As part of the longest study ever conducted on this subject, spanning a 38-year period, Dr. Charles A. Kemper, an expert on bird migration, documented 12,000 pigeon deaths *in a single night* from colliding with a television tower in Minnesota.[31] That may have been an extreme instance, but the problem is persistent and growing. In 2005, wildlife scientist Dr. Albert Manville estimated that up to 50 million birds die this way each year[32]—the US Fish Wildlife Service cites Manville's low-end estimate of four to five million bird deaths, which is still a very large number.[33] For more recent avian tower-collision mortality data for North America, Mexico, and the Caribbean, visit http://www.towerkill.com.

Among the cited causes for this epidemic of avian tower collisions are both the demonstrated effect of magnetic fields on both magnetite systems[34] and cryptochrome systems.[35] By whichever biological mechanism, the data is increasingly clear that the presence of environmentally polluting EM radiation is one of the major causes of tower collisions.

But tower collisions are only part of the story. Low-frequency EM radiation has been shown to alter other aspects of the navigational behavior of birds (which, in turn, impacts avian population). Dr. Alfonso Balmori from Valladolid, Spain, investigated the sparrow population in his town. Once a month between October 2002 and February 2003, Balmori took measurements of MW levels and sparrow populations in 32 different locations around Valladolid. He found that sparrow populations diminish in areas with higher levels of MW radiation. In those areas of the city with the highest levels of such electromagnetic fields, the sparrow population had disappeared entirely. Balmori also noted the reappearance of sparrows in areas where high MW fields had been reduced.

Overall, the sparrow population was migrating from high MW zones (generally toward the city center), to those areas with low levels of microwave radiation (generally on the outskirts).[36] These findings are consistent with what Joris Everaert and Dirk Bauwens reported with the house sparrow population decline from their sample of 150 locations in Belgium.[37] Balmori believes that his results also explain the decline of the house sparrow in England and other parts of Western Europe.[38]

AVIAN REPRODUCTION

Balmori cites the changes in avian reproductive systems as one of the possible effects of EMF exposure (and one of the causes of the decrease in bird populations around Europe), but the effects of EMF exposure on the reproductive function and patterns of birds is not understood quite as well as some of the other impacts investigated.

Dr. Jules B. Youbicier-Simo and fellow researchers presented their findings on this question at the Twentieth Annual Meeting of the Bioelectromagnetics Society in 1998. They had performed a series of three independent laboratory experiments exposing fertilized chicken eggs to microwave radiation directly from cell phones applied from a distance of 10 mm above the eggs for 24 hours at a time. The researchers then compared the mortality rate in these eggs with that in a control group not exposed to cell phone radiation. They found that the average cumulative death rate in the exposed group of eggs (72.3%) was six times higher than in the control group (11.9%). Even more tellingly, the researchers note that the mortalities in the exposed group were essentially restricted to the area immediately around the phone (whereas mortality in the control group followed a more random pattern of distribution).[39] These results are entirely consistent with the population decreases noted by researchers like Balmori and Everaert.

Dr. Kimberly J. Fernie from McGill University in Montreal has published results from a number of studies of the impact of environmental pollutants on the American kestrel. Her research has demonstrated that kestrel exposure to EMF is linked to increases in physical growth,[40] increases in body mass and food intake,[41] suppression of melatonin production,[42] and an increase in oxidative stress (a condition that increases the number of free radicals in the body, thus increasing damage to cells and DNA, and incidence of diseases such as cancer).[43]

Specifically on the question of kestrel reproduction, Fernie has demonstrated numerous effects of EMF exposure on the reproductive success of captive kestrels. In one study, the kestrels were split into two groups. Over a two-year period, one group was exposed to EMF radiation (equivalent to the levels of ELF radiation to which they would normally be exposed from power lines), and the other was a control group, bred in the absence of such radiation. She found that the ELF-exposed eggs were larger and thinner shelled, and while EMF exposure led to an increase in kestrel fertility, the overall hatching success was reduced. The results from this controlled experiment are consistent with the documented findings of reduced reproductive success among kestrels living near power lines.[44]

The science clearly demonstrates that exposure to low-frequency, non-ionizing EMF radiation has impacted many species of birds around the world in a variety of ways, impairing their ability to navigate, altering their habitats, and reducing their populations. Birds, though, are not the only living creatures that exhibit such demonstrable effects of EMF exposure. Among the many others is the honey bee.

COLLAPSE OF THE HONEY BEES

Among the most well-established impacts of EMF exposure on honey bees (as with birds) is the alteration of their navigation patterns. These navigation patterns are vital to human beings because of the role that bees play in food production. Worker bees spend their days flying from their hives to search for food (in the process, pollinating plants). Without their efforts, the hives perish. And, without their efforts, 85% of the more than 200,000 known species of earth's vegetation are unable to reproduce. The impact of the loss of the honey bee would be devastating to humanity; it is estimated that one-third of the food we eat depends directly on their efforts.[45]

This is what makes the phenomenon known as *Colony Collapse Disorder* (CCD) so disturbing. CCD is the name given to the increasingly global pandemic in which honey bees leave their hives and die, diminishing hive populations and in some cases, leading to death of entire colonies. CCD has significantly impacted honey bee populations in North America and Europe. The problem spiked in the United

States between 2006 and 2008. According to a report prepared for the US Congress by the Congressional Research Service, the number of managed honey bee colonies dropped 31.8% in the winter of 2006–7; an additional 35.8% the following year; 28.6% in the 2008–9 winter;[46] and 33.8% between 2009 and 2010.[47] (Keep in mind that those numbers represent national averages; reported losses for some individual bee keepers have been much higher—over 50%.) Between 2006 and 2010, more than 3,000,000 honey bee colonies in the United States disappeared.[48] The numbers are similar in the United Kingdom, per a report from the British Bee Keepers Association, which documented a 30% drop in managed colonies between 2007 and 2008.[49] And again, these numbers are averages that can obscure the scope of the tragedy (one Scottish bee keeper, for example, reported a loss of 80% of his 1,200 hives between 2009 and 2010[50]). At the current rate, experts predict that the honey bee will be extinct in the United States by 2035 and in the UK by 2019.[51] CCD has also been reported in Italy, where according to the European Food Safety Authority, Italy saw a honey bee mortality rate of between 40% and 50% between 2007 and 2008.[52] There are also reports of CCD in Germany and Taiwan.[53]

The precise cause of CCD remains unknown. An emerging scientific consensus indicates that this is most likely due to a variety of stresses—a mix of environmental pollutants (such as pesticides) and biological entities (such as parasites). While it is impossible to prove that any single stimulus is a cause of CCD, the scientific evidence is strong that EMF exposure could be one of the environmental pollutants triggering the phenomenon. EMF exposure in bees has been demonstrated to lead to outcomes consistent with CCD.

HONEY BEE NAVIGATION

Like pigeons and other birds we discussed earlier in this chapter, the navigational ability of honey bees has been scientifically linked to both the magnetite[54] and cryptochrome[55] in their bodies, and thus it's not surprising to discover that EMF interferes with the navigational activities of bees in ways similar to those we have seen demonstrated in bird populations. Some of the results are immediately visually striking—it appears that bees are extremely sensitive to low-frequency electromagnetic radiation.

Honey bee gathering pollen.

A group of researchers in Germany performed a very simple experiment with a cordless phone and a bee hive. They turned on a cordless phone (which is not as powerful as a cell phone) near a hive and investigated whether there was a resulting change in the number of bees that returned to the hive, when compared to a hive without a nearby cordless phone emitting EMF radiation. They performed this test with 16 different hives—8 irradiated and 8 control hives. Of the nonirradiated bees 39.7% returned; only 7.3% of the irradiated bees came back to the hive.[56] These findings are similar to the others published earlier by Dr. Sainuddin Pattazhy in India; except in Pattazhy's experiments, none of the EMF-exposed worker bees returned to the hives at all and the hive completely collapsed.[57]

Another group of researchers from India also linked cell phone radiation to a reduction in hive populations. In this study, the researchers positioned active cell phones around one hive, inactive (dummy) cell phones around a second hive, and no cell phones around a third hive. The active cell phones (emitting 900 MHz microwave radiation) were active for 15 minutes, twice each day, for three months. They noted many effects in the EMF-exposed hive, including reduced flight activity, reduced ability to return to the hive, and a reduced number of worker bees returning to the hive with pollen. The exposed queen laid less than half the eggs of the control queen, the control hive had stored eight times as much honey as the exposed hive, and the population of the exposed hive was reduced by over 70%.58

These experiments demonstrate outcomes—that radiation from cell and cordless phones can lead to reduced hive populations. But they don't tell us how this occurs. Swiss researcher Dr. Daniel Favre performed a series of experiments involving worker piping, the communication among bees that informs them when to swarm and leave the hive, or when the hive is exposed to stresses that disturb the bees. In 83 separate experiments, Favre demonstrated that turning on and using a cell phone near the hive led to a 1,000% increase in worker piping—a 10-fold increase in the signaling to depart the hive.59 In a 2011 interview, Favre explained how this behavior could result in an increase of swarming activity and a decline in hive population.60

PLANTS AND TREES

It is clear that increased levels of EMF radiation in the environment are associated with serious problems in animal life. But, as we recall from Skrunda, it wasn't just human and animal life that was affected, but flora as well. Flora are not only vulnerable to damage from EMF exposure, but according to Dr. Alain Vian of Blaise Pascal University in France, they are even more vulnerable to such damage than are humans and other animals. Plant life, in contrast to animal life, evolved primarily as "surfaces to optimize interaction with the environment." When that environment is polluted, plants have a much greater percentage of cells that interact directly with that pollution.[61]

Consistent with what was seen in the Skrunda pine tree, a study published in the *International Journal of Forestry Research* in 2010 explains that RF exposure has "strong adverse effects" on the tree growth and leaf health of aspen trees. The researchers constructed

Faraday cages (metallic enclosures that can block portions of electromagnetic radiation) around some of the trees, shielding the enclosed trees from normal levels of ambient RF radiation from such sources as radio and television broadcasting; those trees outside the Faraday cages were not protected from the exposures. The aspens shielded from RF radiation displayed significantly increased shoot growth (74% more growth in shoots than the exposed trees) and 60% more leaf area than the exposed trees.[62]

While those exposed aspen and the exposed Skrunda pines both demonstrated reduced size, many trees located near a cell phone tower in Michigan have had the opposite response. As David Reed and his fellow researchers at the Michigan Technological University noted, the red pines near the broadcast antenna have grown larger than those farther away, and the exposed aspen and red maples grew to be thicker than their unexposed counterparts. The red oaks and the paper birches appear to have been unaffected.[63] From these results, Dr. Balmori concludes that the evidence points to EMF having subtle, complex, and varying effects on trees.[64]

Researchers in Germany investigated the cause of a mystery illness that was affecting the health of trees in populated areas of Europe, with symptoms including bark fissures, death of parts of leaves and abnormal growth. It is reported that 70% of trees in Holland demonstrate some or all of these symptoms. Testing the hypothesis that these were symptoms of EMF poisoning, the researchers exposed 20 ash trees to various types of EMF radiation for a period of three months. Those trees exposed to WiFi radiation displayed symptoms consistent with those found in the affected trees.[65]

Just as research indicates that EMF exposure can make humans more susceptible to damage from other environmental pollutants (such as the reduced effectiveness of the blood-brain barrier, discussed in chapter 6, and the increased tolerance to EMF that occurs in cells as a result of repeated EMF exposure discussed in chapter 4), a new study from Sweden demonstrates that some types of EMF can exacerbate damage to trees from other pollutants. These researchers took samples from pine trees located directly underneath 400 kV high-voltage power lines and from other trees located at distances up to several miles away from these lines. The trees under these high-voltage power lines had twice the levels of PCBs, polychlorinated biphenyls. PCBs are synthetic chemicals that had been widely used in electrical equipment until they were banned in 1979 because of their high toxicity to humans; though no longer used in the production of consumer goods, these chemicals remain in the environment. The increased PCB concentrations found under these high-voltage lines are attributed to accumulation of dust particles (which include pollutants such as PCBs) that become charged by the power line, and have an increased tendency to stick to the surface of pine needles.[66]

PLANTS

Like trees, plants suffer damage from exposure to EM radiation, with wide-ranging effects. A study conducted in Romania demonstrated that exposure to 400 MHz EM radiation (between one and eight hours a day, for three weeks), led to a logarithmic decrease in the production of chlorophylls (the green pigment that enables plants to absorb energy from light) in the black locust plant.[67]

Plant cells also have protective cellular mechanisms like those found in animal cells. Researchers from France's Blaise Pascal University demonstrated that EMF radiation triggers the cellular stress response in plants by exposing tomato plants to 900 MHz (equivalent to a cordless phone or a UHF transmission) for 10 minutes.[68] A team in Israel also discovered indicators of cellular stress response in duckweed resulting from exposure to 60- and 100-MHz fields.[69]

That is one of many studies into the effects of EMF exposure on the duckweed plant, in particular. Duckweed is a small, stemless, aquatic flowering plan. A team from the Department of Botany at the University of Zagreb in Croatia performed a series of experiments on the impact of exposing duckweed to 400, 900, and 1900 MHz EMF radiation. These researchers found that many of the exposures altered the growth rate of the duckweed, but noted that "the effects of EMFs strongly depended on the characteristics of the field exposure." For example, exposure to a

900-MHz field strongly inhibited growth, whereas a 400-MHz field did not.[70]

Two years later, the same team performed similar experiments on duckweed and noted that exposure to EMF between 300 MHz and 300 GHz increased oxidative stress (which is a sign of the plant's inability to cope with and process toxins) as well as the production of antioxidative enzymes.[71] Similar alterations to antioxidant behaviors have also been found in tobacco. Experiments conducted at Tarbiat Modares University in Iran, published in 2007, found that exposing cells of the tobacco plant to low-frequency EMF for five hours a day for five days, led to changes within cells that could damage the antioxidant defense systems of cells.[72]

PLANT GROWTH

With all of these changes in plants at the cellular level resulting from EM exposure, we might expect to see changes in plant growth and yield (as we do in trees). And this is what we find. A team at the Department of Environmental Research in Austria investigated the effect of EMF exposure from high-voltage power lines on the output from wheat fields over a five-year period. The wheat closest to the tower had EMF radiation strength approximately 11 times greater than the wheat that was farthest away, and they found that the grain harvest from the least exposed group was 7% higher than from the wheat closest to the transmission tower.[73]

Much of the research into EMF and plant growth has been performed on seeds and seedlings, with very different results depending on the strength, frequency, and duration of exposure, as well as the type of seed.[74] Looking at the effects of cell phone radiation on mung bean seedlings, a team in India exposed mung beans to cell phone radiation for thirty minutes, one hour, two hours, and four hours per day. These researchers found that the exposed seedlings had significantly reduced length and weight as well as reduced levels of proteins and carbohydrates. From this data, they concluded that exposure to cell phone EMF impairs early-stage growth in mung beans.[75] A separate study into mung beans conducted in Taiwan found that certain frequencies of ELF radiation (30 Hz, 40 Hz, and especially 50 Hz) also inhibited early-stage mung bean growth, but other frequencies (20 Hz and 60 Hz) actually enhanced its development.[76]

IMPACT ON THE ENVIRONMENT

As we proceed with our evaluation of the impact of electromagnetic fields, it is important to remember that humans are not the only life forms directly affected by the increase of EM pollution in the environment. We are also affected indirectly as part of an ecosystem where damage to any part affects the whole negatively in some way. EMF radiation goes everywhere and interacts with all forms of life. This is understandable given that EMF damages DNA, and DNA is found throughout all organic life.

The literature shows broad-ranging, scientifically demonstrated impacts of EMF pollution on plants and animals. As with humans, there appears to be no single level for biological responses to EMF exposure in plants and animal life—indeed, many of these studies seem to indicate there are an even greater variety of health effects on plants than on animals.

This science provides some telling clues as to the extent of the effect on nature stemming from man-made RF/MW and ELF radiation from wireless telecommunications, transmission towers, and high-voltage power lines—all of which are in near-continuous operation in their environments. In addition to loss of territory through habitat deterioration, many species of animal and plant life suffer long-term health effects and reproductive consequences from EMF exposure.

Given the well-documented health effects of EMF exposure, why isn't a greater alarm being sounded? That's the subject of the next chapter.

Chapter 8

THE BUSINESS OF EMF SCIENCE

In the last few chapters, we've reviewed the scientific literature that has investigated the health effects of EMF exposure. Today, the body of science informing us about the biological and health effects resulting from EMF exposure is much larger and more rigorous than it was when the issue first became public a quarter century ago. While the picture is by no means complete and we have many more questions to pursue, the science clearly demonstrates that non-ionizing EMF radiation does harm humans and other forms of life, causing disease and other health disorders.

So, why do we often hear that this science is inconclusive? The answer is that while many studies show negative effects, many others show no effect at all. If the hazards are real, why do so many studies demonstrate no health effects? To answer this question, we shall delve into the history of EMF science and biology in the 20th century.

DR. ZORY GLASER'S 1971 *BIBLIOGRAPHY*

Until the middle of the 20th century, research into the effect of radio frequency and microwave radiation on humans focused on potential medical and therapeutic applications, such as the ability to heat human tissue. With World War II, this changed. The introduction of technologies such as radar, that relied on and emitted large amounts of RF and MW radiation, revealed a new potential for this technology. Following that period, research on the biological effects of microwave radiation shifted from a medical pursuit to a military-industrial pursuit.[1]

By 1971, radar and similar technology had been in wide use by the military for decades. Throughout this time, there had been concerns of biological effects—baldness and sterility chief among them—in those exposed to RF from radar technologies.[2] Increasingly concerned with the possible negative health effects on its personnel, the US Navy wanted to understand precisely what the known science of the time indicated. So in 1971, the Navy assigned Zory Glaser, a young PhD working at the Naval Medical Research Institute, the task of creating an inventory of the science of biological effects resulting from RF exposure.[3]

Dr. Glaser reviewed and cited *over 3,000* scientific studies on the biological effects of exposure to EMF in the first version of what he referred to as his *RF/Microwave Bioeffects Bibliography*.[4] Many of these experiments had been conducted by the Soviet military (which was also very interested in the same questions). Impressed with these results, the Navy asked Dr. Glaser to maintain his bibliography over time, which he did over the following decades, ultimately accumulating data from over 6,000 separate studies.[5]

From Glaser's work, we know that the US and former-Soviet military have been aware of the potential for negative health effects resulting from exposure to RF and MW radiation—even at very low power levels—for over 50 years. Work published in 1926 by a surgeon with the US Public Health Service demonstrated lethal results from exposure to ultrashortwave EMF in mice. The surgeon attributed the mortality to mechanisms other than the heat response, concluding that exposure to EM radiation resulted in nonthermal health effects in the animals.[6] Two years later, German researchers reported fatalities in mice, rats, and flies from similar short-wave exposures.[7] In the early 1930s, a zoologist at the University of Pennsylvania noted the effect of EM radiation on wasps and frogs, concluding, "it is evident here that high frequency and heat are by no means synonymous and that though the electrostatic field carries with it potentialities for internal heat as a by-product, there is at the same time another and little understood reaction."[8]

In 1948, two groups of researchers, working independently, both noted nonthermal effects resulting from EM radiation exposure. Scientists at the Mayo Clinic noted the incidence of cataracts in dogs following exposure to microwave radiation, and researchers at the University of Iowa noted that exposure to microwaves resulted in cataracts in rabbits and dogs, and "testicular degradation" in rats.[9]

In the early 1950s, a physician named John T. McLaughlin was working at Hughes Aircraft Corporation in California, where he noted between 75 and 100 cases of a form of internal bleeding known as *purpura hemorrhagica* among 6,000 workers—an unusually high incidence. McLaughlin suspected microwave radiation exposure as the cause, and in the course of his investigation, he also noted several cases of cataracts and headaches among those working near sources of microwave radiation.[10]

Soviet scientists recognized that EMF at frequencies between 30 MHz and 300 GHz could affect the human circulatory system (altering heart rate and blood pressure) and nervous system, even at levels too weak to produce thermal effects. Further, these scientists found that symptoms depended on the length of employment and degree of exposure, demonstrating a clear dose-response relationship.

The Soviets were so interested in the health effects of microwave radiation, that they weaponized it. In one prominent case, it subsequently became known that the Soviets bombarded the US embassy in Moscow with microwave radiation since the 1950s until the mid-1980s—"at the same time that they were pursuing a very active research program on low-level, chronic effects."[11] The exposures resulted in reports of "inexplicable health problems" among the embassy personnel; many believe that the death of former US ambassador Walter Stoessel from leukemia at the age of 66 resulted from his exposure to Soviet microwaves during his tenure in Moscow from 1974 to 1976.[12]

These are just a handful of the scores of studies covered in Glaser's bibliography. Still, results such as these from scores of studies did not set off alarm bells in the general public, and understandably so, given that the public still did not commonly have the devices that produced RF and MW radiation, such as the cell and cordless phones, WiFi networks, and microwave ovens that are ubiquitous today. While there were some exceptions (such as those residents near Skrunda, Latvia, exposed to high levels of RF from the nearby military base discussed earlier), in general the public had no exposure to these frequency ranges of damaging non-ionizing radiation. Of course, that started to change in the 1980s, with the introduction of cell- and cordless-phone technologies.

THE EPA AND EMF

By 1989, cell phone use was still quite limited in the United States, with only approximately 1.4% of the population having cell phone access.[13] Still, public concern had grown to the point where the US Congress's Office of Technology Assessment issued a paper calling for Americans to practice "prudent avoidance" with EMF exposure in the home.[14] In 1989, investigative science writer Paul Brodeur authored the first popular national article highlighting the public health threat of power-line-frequency EMF and the lack of government action on the subject in a three-part series for the *New Yorker*.[15] Awareness of the potential dangers of EMF exposure was starting to enter the public sphere. It was in this context that the Environmental Protection Agency (EPA) initiated a review of the known science on the biological effects of exposure to RF/MW radiation, with the goal of releasing an official summary.

In one of the drafts of this report, released in March 1990, the EPA's Office of Health and Environmental Assessment (OHEA), then headed by Dr. Robert McGaughy, recommended that EMFs be formally designated as known "probable human carcinogens" and that RF/MW radiation in particular should be considered a "possible human carcinogen" (along with other class B carcinogens such as DDT, PCBs, and formaldehyde).[16]

Emphasizing the significance of this wording, the *New York Times* reported on the draft. The 1990 article entitled "Study Says Electrical Fields Could Be Linked to Cancer" quoted then-OHEA director, Dr. William Farland, who noted an important shift: "Over the past few years, more and more people have begun to say there does seem to be something there, that we need to do more work, whereas before we were saying that it was not worth pursuing. This is an important step in getting more research done."[17]

The following year, however, this language was stripped from the draft of the report by the EPA Science Advisory Board and the Nonionizing Electric and Magnetic Fields Subcommittee of the Radiation Advisory Committee.[18] In its place was added the following:

> At this time such a characterization regarding the link between cancer and exposure to EMFs is not appropriate because the basic nature of the interaction between EMFs and biological processes leading to cancer is not understood.[19]

Strangely, the same page stated that several studies suggested a "causal link" between exposure to 60 Hz EMF and leukemia and lymphoma in children and workers.[20] Despite that inclusion, the most explosive elements of the EPA's initial findings—what Dr. Farland had explained as "an important step in getting more research done"—had been scrubbed from the report.

Why was this specific language around carcinogenicity removed from the draft?

The EPA explained that use of the term *carcinogen* was "not appropriate" until better data existed demonstrating this link (what levels of exposure, at which frequencies of EMF, for what duration, caused which specific health outcomes). In short, the EPA explained that while it had some data indicating the health risks from cell phone radiation, it needed more specific proof before labeling EMF as carcinogenic.

As an external observer, it is impossible to say what occurred behind the scenes to trigger the removal of this potentially explosive language from its report. There can be no doubt, however, that the 1990 EPA draft coincided with an aggressive effort from the wireless industry to refute any such potential associations between cell phone radiation and negative health outcomes—particularly cancer—in humans. After all, the government was considering labeling its core products as carcinogenic.

By the mid-1990s, the wireless industry's effort to defang the EPA was in full swing. In 1995, Douglas Bannerman of the National Electrical Manufacturers Association in Arlington, Virginia, argued that "we should not have individual agencies popping up and giving their own risk assessments."[21] The US Senate agreed, cutting $350,000 from the EPA's budget because the Senate Appropriations Committee "believes [the] EPA should not engage in EMF activities."[22] By 1996, due to "budgetary uncertainties," McGaughy explained that "the report will not come out in the foreseeable future."[23] The EPA never issued its report; subsequent federal reviews of EMF research have been conducted by other agencies, including the National Institutes of Health (NIH) and the Centers for Disease Control (CDC),[24] both of which conclude that there is no scientifically demonstrated risk from EMF exposure.

The 1990 draft report from the EPA may have set off a firestorm, but its effects were limited. By and large, the public remained unaware of the EMF issue. That started to change on January 21, 1993.

EMF GOES MAINSTREAM

That evening, David Reynard, a Florida businessman from Madeira Beach, appeared on CNN's *Larry King Live* to announce his lawsuit against the cellular phone industry. He explained that his 33-year-old wife, Susan, had died of a brain tumor seven months earlier. Her fatal illness, he alleged, was directly linked to her cell phone use. Just four years earlier, Susan had started using the wireless device when she became pregnant, after Reynard had given her his portable phone. In spite of many unusual complications during her pregnancy, Susan eventually gave birth to a healthy baby, six weeks premature. That's when she underwent an MRI, and the tumor was detected.

> KING: When did you start to think, "This has something to do with the cellular phone?"
>
> REYNARD: I think when I saw the first MRI and saw the location of the tumor. It appeared that it was in the location directly next to the antenna, and the tumor seemed to be growing inward from that direction.

Reynard made a compelling presentation of the X-rays, which showed the tumor right next to where his wife had held the phone against her head. Reynard's legal action against the cell phone manufacturer, NEC, launched their mission to raise public awareness about their concerns. "I don't think [people] realize . . . that these are microwave devices," David said. He then made a bold comparison between cell phones and cigarettes, expressing his strong belief that the wireless devices should also carry FDA health warning labels on the packaging.

King pointed out that the cellular phone was the number one Christmas gift that year, which saw over 16 million Americans with cell phone subscriptions,[25] up from just 3.5 million four years earlier.

A PUBLIC RELATIONS NIGHTMARE

The next day, the Reynard lawsuit headlines splashed across front pages nationwide. The AP announced that there was a "widower on a mission against cellular phones."[26] "Telephone firms fight cancerous connection," explained Reuters.[27] "Woman's Death Fuels Phone Fears, Cancer Scare Rocks Cellular Industry," wrote Florida's *Palm Beach Post*.[28] The *Sarasota Herald-Tribune* headlined their coverage of the Reynard case with "Cellular Phone Scare Hits Stock Markets." The article went on to say that "potential customers, who had been signing up at a rate of more than 7,000 a day, are now asking [cell phone] dealers pointed questions or delaying purchases."[29]

The Reynard case was dismissed by the Florida Circuit Court in 1995 on the basis of insufficient evidence. Yet, having a face to put to this potential hazard sparked the public's concern. Telecommunications stocks took a massive blow. Shares of Motorola plunged 20% by January 30th compared to the price the day before Reynard went on Larry King. McCaw Cellular Communications (another large cellular provider at the time) plunged 15% in the same timeframe.[30] There was no way around it: the cellular phone industry had a major public-relations problem on its hands.

The wireless industry immediately went into crisis-management mode. Thomas Wheeler, then president of the Washington, D.C.-based wireless trade group known today as *CTIA—The Wireless Association*, promptly called a news conference to reassure the public. He issued a statement saying that more than 10,000 studies over 40 years showed no evidence linking cell phones and health hazards. The problem was that Wheeler could not actually produce any scientific studies to back up his claims. The studies to which he referred only assessed microwave ovens.

At the same time, US Congressman Edward Markey, a Democrat from Massachusetts, asked the US General Accounting Office (GAO) to research whether the phones posed a health risk. He convened a spur-of-the-moment telecommunications meeting to hear testimony about the safety of cell phones. It became clear that the wireless industry had not been required to do any premarket testing or postmarket surveillance on the health effects of their product. Even so, the FDA released a statement announcing that "there is no proof that cellular telephones are harmful"[31] but nevertheless encouraged consumers to limit the time they spend talking on the phones if they were concerned.

Feeling the pressure, Wheeler pledged $25 million toward a research initiative to dispel the public's fears. The Wireless Technology Research (WTR) research program, as it came to be called, was set up to conduct the studies, with oversight from the FDA. Dr. George Carlo, a well-known epidemiologist and medical scientist, was tapped to lead the effort. He ultimately assembled a team of 200 scientific experts charged with examining the potential dangers of cell phones. And, by his own account, Carlo was initially successful in this position, boasting in 1994 that "a concerted industry response succeeded in blunting unsubstantiated allegations about a link to brain cancer in early 1993"[32] (referring to the Reynard case). (To get an idea of the politics of the EMF issue, note that the same Thomas Wheeler was nominated by President Obama in 2013 to head the FDA.)

CARLO'S RESULTS

Five years later, in February 1999, Carlo released the WTR results to the public, and his findings stunned the very industry that hired him.

According to his report presented to the annual CTIA convention in California, Carlo had found the presence of micronuclei (DNA fragments) in the blood, indicating that the radiation from mobile phones had caused irreparable DNA damage in cells. (As noted earlier, the relationship between micronuclei and cancer is so strong that physicians around the world test for the presence of micronuclei in the circulation to identify patients likely to develop cancer.)

These conclusions were not well received by the wireless industry. The CTIA responded by discontinuing Carlo's funding and trying to discredit him and his six years of research. Carlo has subsequently become a public health activist on the subject of EMF pollution. In 2010 he said, "today, I sit here extremely frustrated and concerned that appropriate steps have not been taken by the wireless industry to protect consumers." He continued:

> Indications are that some segments of the industry have ignored the scientific findings suggesting potential health effects, have repeatedly and falsely claimed that wireless phones are safe for all consumers including children, and have created an illusion of responsible follow-up by calling for and supporting more research.[33]

Unfortunately, the response of industry to Carlo and his work is not anomalous. Quite the opposite, it is indicative of a larger pattern, demonstrating some of the core ways in which industry interests use the business of science to manipulate the scientific data that is produced and how it is framed, disseminated, and interpreted. The type of science that researches the biological and health effects of EMF is expensive, requiring significant funding—money that comes from the industry that trades in the product being researched. By controlling the funding of the science, industry significantly influences the publicly accessible data on this vital public health issue.

Carlo is just one prominent example; another is Dr. Henry Lai.

WAR-GAMING DR. HENRY LAI

Today, the work of Drs. Henry Lai and Narendra Singh, demonstrating DNA damage from non-ionizing EMF (discussed in chapter 4) is almost two decades old, and the intervening years have seen multiple peer-reviewed studies that have supported its results. At the time, however, when common wisdom held that non-ionizing EMF was biologically benign, the news reverberated throughout the still-nascent cell phone industry. Ironically, as Lai explains, they weren't looking to set off any controversy. Indeed, Lai and Singh weren't even considering cell phones when they executed this research—instead, they were focused on the health effects of RF exposure from radar.

Even so, attempts were made to have Lai discredited, defunded, and fired. In retrospect, we know this effort was intentional. An internal memo leaked from inside Motorola (at the time, the world's second largest cell phone maker) and published in *Microwave News* reads that Motorola executives believed they had "sufficiently war-gamed" Lai and his study.

After the release of the DNA strand-break study, someone called the NIH (from whom Lai received research funding) anonymously to report that Lai was using the funds to execute research outside of the approved scope of the grant. The NIH investigated and eventually dismissed these allegations.[34] Despite the harassment, Lai wanted to continue his research. He applied for and received funding under CTIA's WTR program. But he found the conditions of the grant and the way in which it was managed so unusual and disheartening that he expressed his concerns in a public letter published in *Microwave News*, where he lamented the "consistent pattern of chaotic corruption and deception" in the WTR research program.[35]

In response, the CTIA sent multiple letters to the president of the University of Washington (where Lai and Singh were on faculty), demanding that both be fired.[36] (I was among those who sent letters in support of Lai to the University of Washington at the time.) They retained their jobs but proceeded to see their research funding dry up. In the late 1990s, Lai began to search for European funding sources, later explaining that the United States had been on "the cutting edge of this whole area for the last 30 years. [But] right now, we're the Third World country. We're not doing research at all."[37]

While discrediting and defunding Lai was part of Motorola's strategy, there were other important aspects to their plan. Another part of Motorola's "war game" against Lai and Singh involved Dr. Jerry Phillips, a researcher based in the laboratories of Dr. Ross Adey in Loma Linda, California.

JERRY PHILLIPS AND MOTOROLA

In the early 1990s, concerned with the implications of emerging research indicating potential health effects of exposure to EMF from cell phones, Motorola had begun sponsoring Phillips to perform research under their auspices. Phillips had spent years investigating the potential health effects of 60Hz-fields associated with power lines and electrical wiring under a series of grants provided by the US Department of Energy. With this new round of research sponsorship, Phillips expanded his inquiries further up the EM spectrum.

This funding was already in place when Lai and Singh published their groundbreaking results and Motorola contacted Phillips, asking him (according to Phillips) to "put a spin on the study" that would be more favorable to Motorola.[38] Phillips declined but did offer to conduct a similar trial, to see if he could replicate Lai and Singh's results. Motorola provided funding, and the work began. To ensure that his team was able to actually conduct a study that properly replicated the work of Lai and Singh, Phillips began this effort by sending two of his research assistants to work in Lai's laboratory and learn his techniques.

Initially, Phillips and Motorola had "very cordial" relations. "But only until we started getting data that they didn't like," explains Phillips[39]—which was almost immediately. The first set of data Phillips prepared for publication appeared to demonstrate a biological effect of microwave exposure on proto-oncogenes (a normal gene that has the potential to become carcinogenic)—similar to effects Phillips had observed with his earlier work with 60 Hz EMF. Phillips's draft of the paper noted the potential health effects indicated by the changes in proto-oncogenes. Motorola's head of research, Mays Swicord (who had come to Motorola from his previous job at the US Food and Drug Administration[40]), contacted Phillips and expressed his desire for that language to be changed. Phillips refused. Yet, when Phillips saw the published paper in *Bioelectromagnetics*, the language had been altered, per Motorola's preferences conveyed by Swicord to express doubt as to the physiological implications of the findings. Motorola had edited Phillips's document, without his consent, to present the data with a less damaging analysis.

Shortly thereafter, Phillips requested approval from Motorola to present results from the Motorola-funded research at the annual conference of the Bioelectromagnetics Society in Victoria, British Columbia. The data concerned rates of DNA damage in animals exposed to RF. Motorola informed Phillips that he would have to change many of the statements for the presentation. Motorola "didn't want, for instance, any mention of damage to DNA and radio frequency fields in the same abstract."[41] Ironically, Phillips's data demonstrated a *decrease* in the rate of damage to DNA resulting from RF exposure. But as Adey (Phillips's supervisor) later explained, Motorola was uncomfortable with any "evidence that mobile phones appeared to be having a biological effect."[42] Data that demonstrated any health effect at all was of concern to Motorola.

THE BREAKING POINT

By 1997, Phillips had completed this research. He and his team had found that exposure to RF radiation had increased DNA damage in some instances and decreased it in others. He submitted these findings to Motorola. Motorola's Swicord contacted Phillips to discuss the apparent inconsistency in the data. While these results may appear confusing or contradictory, Phillips explained that these results do make sense, because "if you produce a little bit of DNA damage, you are stimulating the repair mechanisms and you could actually see a net decrease because the repair will be done. However, if you overwhelm the repair mechanism, then you could see an increase" in DNA damage.[43] This could be why exposure to a relatively weak EM field can yield such different outcomes compared to larger or more extended exposure.

Despite the explanation, Swicord pushed for Phillips to continue his study, offering him more money to produce more data, before publishing any conclusions. Phillips refused. "I said, no. The study's done. I've been doing research for over 25 years. I know when a study is done. I'm going to go ahead and publish the work."[44] Adey strongly encouraged Phillips to give Motorola what it wanted, indicating that not doing so would harm Phillips's career. Again, Phillips resisted, and he published the findings in *Bioelectrochemistry and Bioenergetics* in November 1998.

As Adey had warned, this proved to be the end of Phillips's research funding from Motorola. Unfortunately, this coincided with the end of his funding from the Department of Energy, leaving Phillips with no sponsorship to run his laboratory. Phillips and his wife opted to leave research and moved to Colorado, where he is director of the Health Science Learning Center at the University of Colorado in Colorado Springs. Later, in 2009, Phillips, Lai, and Singh together published a joint paper reviewing effects of EMF on DNA in a special issue of the journal *Pathophysiology*.[45]

BETTER RESULTS

After Motorola proved unsuccessful at suppressing publication of Phillips's results, their funding shifted to other researchers who provided more comforting data. As a Motorola representative explained in a public statement issued on Colorado's KGNU radio, "Motorola commissioned a separate laboratory to follow up on the results published by Dr. Phillips. That and other studies have failed to confirm his conclusions."

One of the laboratories that received research sponsorship from Motorola was Battelle Pacific Northwest National Laboratory in Richland, Washington (whose chief scientist of the Health Division, Dr. Thomas

Tenforde, said in 1996 that "there are limits to what one can consider for the sake of safety without going back to the Dark Ages"[46]). Battelle was unable to reproduce the results reported by Phillips or Lai and Singh.[47] Battelle has subsequently produced other studies supporting the claims of the wireless industry, including research challenging the validity of the link between EMF exposure and breast cancer.[48] Battelle also produced data disputing the results presented by my colleague Dr. Reba Goodman (discussed in chapter 1) on the effects of EMF exposure on gene expression.[49]

When trying to understand seemingly contradictory results from competing scientific studies, it is important to know that one needn't falsify data in order to misrepresent truth. In the case of Battelle's claims against Dr. Goodman, for example, it turns out that Battelle's "replication" of Goodman's study, did not actually replicate her study. The specific types of cells researched in both studies (known as HL60 cells) were from two different suppliers and had very different growth characteristics. A seemingly inconsequential and easily overlooked detail, such as the source of the cells, accounts for the difference in outcomes between the studies. As a result, the data from both studies were correctly reported, but Dr. Goodman's conclusions were based on functioning cells. Battelle's conclusions were based on cells with greatly impaired function, and therefore not valid, although coincident with the preferences of their funding source.

As this example demonstrates, it is exceedingly easy to alter or tweak the design of a scientific study in seemingly minor ways that generate significant differences in outcomes. These experiments are quite complicated, and there are a tremendous number of variables involved. The choice of cells to study, the EM frequency, the duration of exposure, and the cumulative exposure from all sources are just a few of the factors. Then there are the procedural considerations, such as precision of scientific techniques and the creation of stable control groups.

And these are just the variables that go into the study. Interpreting the data presents an entirely different set of complexities, as we saw with Phillips's attempt to explain DNA damage rates to Swicord.

Recall that, at first glance, Phillips's data appeared to demonstrate inconclusive results—some exposures increased DNA damage, some decreased DNA damage—even though there is a perfectly valid scientific explanation. Given such complexities, even minor alterations to assumptions or the execution of the study can lead to seeming contradictions.

SEEMING CONTRADICTIONS

Sometimes, these apparent contradictions result from seemingly trivial differences in approach. For example, in 1997, a group in Australia studying mice exposed to cell phone radiation reported a significant increase in the incidence of lymphomas—a type of cancer of the blood.[50] Then, a few years later, another group of investigators attempted to replicate the study. Their results indicated that "there was no significant difference in the incidence of lymphomas between exposed and sham-exposed groups at any of the exposure levels."[51]

At first glance, it seems that these results are contradictory. But on closer inspection, we see that this is not actually the case. In the first study, the mice were not handled by the technicians; in the second study, which supposedly replicated the conditions of the first, the researchers did handle the mice. It has long been known that manual handling of mice is a form of stress that contributes to disease. Thus, in the second study *both* the exposed mice and the control mice were exposed to environmental stress, and as a result, both groups showed increased cancer formation. As a result of the increased cancer in the "control" group, the increase in the EMF group turned out to be below the level of significance, even though it was larger. The second study may have attempted to replicate the first, but in actuality it did not. The introduction of handling into the procedure might seem to be an inconsequential factor, but in this case it proved critical in determining the outcomcs.

Other times, however, the contradictory results of such studies are not accidental, but intentional. As Phillips explains, "there are some scientists who know that you can design an experiment in this area to produce any sort of effect you want . . . If I want to set up a study that's guaranteed to show no effect, I can do it." And this is what Motorola proceeded to do. Phillips continues:

> Motorola has purchased results. I mean, I know laboratories that they've funded to the tune of many millions of dollars, and these labs have produced one study after another after another that say, Lai and Singh were wrong. Look, this other lab was wrong. Look, Phillips was wrong. Everybody is wrong, except these other people hired by Motorola.[52]

DRS. ROTI ROTI AND MALYAPA

Another of the Motorola-funded studies is the work of Drs. Joseph Roti Roti and Robert Malyapa of Washington University in St. Louis. In 1997, Roti Roti and Malyapa published their findings in *Radiation Research* from a study sponsored by Motorola. The results demonstrate that exposure to 835 MHz, 847 MHz, and 2,450 MHz MW radiation did not increase DNA strand breaks in cultured cells. (Separately, they claim to have been unsuccessful at identifying any changes within the brains of rats exposed to MW radiation.[53]) Not surprisingly, Lai and Singh dispute these claims.

As you'll recall, when Phillips wanted to replicate Lai's work to see if he could reproduce their results, his first step was to send his researchers up to Lai's laboratory in Washington to learn all of the details of the study. This is a desirable protocol for increasing the accuracy of attempts to replicate scientific experiments. Roti Roti did not do this (neither did Battelle). In his study, Roti Roti used a different variant of the *comet assay* technique (to measure the rate of DNA strand break) than that used by Lai and Singh. Both Lai and Singh explain that the technique used by Roti Roti cannot actually detect DNA damage from RF radiation. But in any case, one cannot cite the Roti Roti study as evidence to doubt Lai's work—because the Roti Roti study did not actually reproduce the same conditions and techniques as Lai's studies.

By design, Roti Roti and Malyapa were testing something else. And, regardless of the intent of the researchers, they produced results that were in accord with their funder's, Motorola's, preferences.

The Roti Roti study, as well as the conflicting studies regarding lymphomas in mice exposed to cell phone radiation, speak to the importance of how a study is designed—choosing which type of experimental (laboratory) data to analyze and how to collect it. This is even more important in epidemiology studies. An example of how study design can influence the results of an epidemiology study can be found in a well-known and frequently cited 2006 study conducted in Denmark examining the health effects of cell phone use.

2006 DANISH STUDY

This study was massive in scope. Due to unique characteristics of the Danish health system and cell phone networks, the research encompassed the entire population, tracking cell phone usage of 420,095 subjects who began using the mobile devices between 1982 and 1995. For the subjects with the longest history of cell phone use, the study included 21 years of health history following their first exposures to cell phone radiation. This is an extremely long period for these types of studies, which is important given that it can take between 10 and 25 years for certain types of cancer to form.

From their analysis of this extensive data set, the researchers concluded:

> We found no evidence for an association between tumor risk and cellular telephone use among either short-term or long-term users. Moreover, the [high degree of statistical accuracy of our data] provide evidence that any large association of risk of cancer and cellular telephone use can be excluded.[54]

Given the scope of this study and the high degree of confidence with which the researchers report their unambiguous analysis, one would think that these results cast significant doubt on the claims of those who believe that RF/MW radiation is carcinogenic. Unfortunately, according to Dr. Carlo (who, as you'll recall from earlier in this chapter, previously headed the CTIA's WTR program and administered $25 million of industry funding for scientific studies into RF/MW radiation), this is not the case. "This study, funded by the telecommunications industry, was clearly created in order to produce a positive, low-risk finding." As Carlo explains, this "Danish Cohort Study was epidemiologically designed to produce a pre-ordained positive outcome."[55]

FAULTY BY DESIGN

A more detailed review of the findings from the Danish study reveals that the researchers defined a cell phone user as someone who used a cell phone at least one time per week. As Carlo explains, "finding a cell phone related cancer risk among this group would be akin to identifying excess lung cancer risk among people who smoked one cigarette a week—similar to finding a needle in a haystack." Neither does the analysis account for actual time spent on the phone, which was significantly lower in the 1980s and 1990s than it is today, or even when the study was published in 2006. Across the period of the study, the 420,095 participants averaged 17 to 23 minutes *per week* of cell phone usage (by way of comparison, the average American today spends more time than that on cell phone calls *each day*). Even among vigorous opponents of cell phone use, few would be concerned about the minimal periods of use that appeared in the Danish study.

Not included in 420,095 subjects were over 200,000 cell phone subscribers who were dropped because they were corporate customers for whom individual information was not available. This not only represents a sizable loss of approximately 30% of potential study subjects, it also represents the loss of the group that is reputed to be among the heaviest users (particularly in the 1980s and 1990s, when cell phones were still exotic and very expensive devices). Even more significantly altering the nature of the results, these corporate users were categorized in the control group. "In other words, these heaviest users were treated as if they did not use cell phones. In his report on the May IARC meeting, Baan (a visiting scientist at the IARC) wrote that this 'could have resulted in considerable misclassification in exposure assessment.' That's just a smart way of saying that the study has a good measure of bias."[56]

Carlo notes other flaws in his critique of the study, as does Dr. Lennart Hardell in his paper published in the Oxford Journal of Medicine.[57] For instance, users of cordless phones (exposed to RF/MW radiation from their phones), who did not use cell phones, were classified in this study as "nonusers." Still, these fundamental flaws in the design of the study did not stop its wide reporting around the world. In the United States, CBS News covered the study with the headline "No Cancer Risk Seen with Cell Phones;"[58] USA Today concluded "A huge study from Denmark offers the latest reassurance that cellphones don't trigger cancer;"[59] and the San Jose Mercury News led their coverage with "Will your cell phone cause cancer? No, according to a study by Danish researchers who tracked 420,000 cell phone users."[60]

FUNDING BIAS

Private industry, including firms like Motorola, and the trade organizations that represent them and others, like CTIA, have now funded a significant amount of research into the question of EMF. Indeed, private sources are now essentially the only type of funding for these types of studies in the United States (the situation is different in Europe). Dr. Gene Sobel, who concluded that there is "strong epidemiological evidence" for the link between EMF and Alzheimer's, explains that "it's next to impossible to get money to do these studies."[61] And, as we've seen, the private funding has tended to end up with those researchers who produce results favorable to the profits of the wireless industry. As Lai says, "the mechanism is funding . . . You don't bite the hand that feeds you. The pressure is very impressive."[62]

As a result, the science on the biological effects of exposure to EM radiation has been subject to significant *funding bias*, or the tendency of outcomes from studies to align with the interests of those funding the studies.

Numerous studies have been performed analyzing the impact of funding bias in multiple arenas of science and public health. Dr. Christina Turner reviewed 91 papers investigating tobacco and cognitive performance, concluding that "scientists acknowledging tobacco industry support reported typically that nicotine or smoking improved cognitive performance while researchers not reporting the financial support of the tobacco industry were more nearly split on their conclusions."[63]

On the controversy over the chemical bisphenol A (BPA) in plastic products, the *Washington Post* reported that "more than 90 percent of the 100-plus government-funded studies performed by independent scientists found health effects from low doses of BPA, while none of the fewer than two dozen chemical-industry-funded studies did."[64] And a review of studies on drug trials in the pharmaceutical industry noted that "company-funded trials are four times more likely to find evidence in favor of the trial drug than studies funded by other sponsors . . . As a result, it is largely impossible to reliably assess the benefit and harm of medical drugs on the basis of published trials."[65]

This same funding-bias effect—aligning the interests of funding sources with the scientific outcomes produced by the sponsored researchers—has been repeatedly demonstrated in EMF science. Since 1990, Lai has been tracking the studies of the health effects of RF radiation on humans published around the world. He has hundreds of such studies in his database. Approximately 30% of the studies are funded by the wireless industry and 70% are funded by other sources that are presumably more independent. Of the industry-funded studies, 27% demonstrated a biological effect in humans resulting from RF exposure; whereas independently funded studies found such effects in 68% of the studies.[66] As Lai explains, "a lot of the studies that are done right now are done purely as PR tools for the industry."[67]

Similar results were presented by Dr. Anke Huss's review of 59 studies related to the health effects of mobile phone radiation. Studies funded by industry were nine times more likely to demonstrate no health effect than those studies funded by public or charitable sources.[68] She found that 82% of research funded by public agencies (such as governments) and 71% of research funded jointly by industry and the public reported health effects resulting from RF exposure. Of the industry-funded studies, only 33% demonstrated such a link.[69] As Huss concluded of the science on bioeffects of EMF exposure, "studies funded exclusively by industry reported the largest number of outcomes, but were least likely to report a statistically significant result."[70] Later, in 2010, Joel Moskowitz, from the University of California at Berkeley, reviewed 23 case-controlled studies examining the potential link between cell phone use and the risk for tumors, concluding that "among the 10 higher quality studies, we found a harmful association between phone use and tumor risk. The lower quality studies, which failed to meet scientific best practices, were primarily industry funded."[71]

These reviews reveal the success of another of the wireless industry's tactics in its scientific battle to defend its profits. After attacking and defunding scientists who publish results suggesting negative health effects from EMF, and shifting funding to other researchers who produce data more in line with maintaining profits, the industry then simply counts up the studies and presents the issue to the public as a simple scoreboard. As Joe Farren, CTIA's director of public affairs, explains, "any official precautionary measures need to be based on the science. The majority of studies have shown there are no health effects."[72] In other words, we have more science on our side; therefore, cell phones are safe. This well-funded messaging has influenced the public discourse, as we see in an April 2012 article in UK's *Telegraph*, which reassuringly explained:

> Two years ago the INTERPHONE study [discussed in the next chapter] reported that the heaviest users could be at a 40 per cent increased risk of developing glioma, a common type of brain cancer. *Most studies have found no such association though* [emphasis added].[73]

Some reviews of EMF health impacts by more explicit conflicts-of-interest can be harder to uncover. As we already saw with the example of Mays Swicord, who worked at both the US FDA and Motorola, there is a revolving door between the wireless industry and government regulatory agencies. (Swicord was at Motorola when he was elected president of the Bioelectromagnetics Society and advocated ending research on EMF as a potential health hazard.) Sometimes, however, the door is not so much revolving as nonexistent.

CONFLICTS OF INTEREST

Dr. Anders Ahlbom is a professor of epidemiology at the Karolinska Institute in Sweden. He was, until recently, a highly influential and respected scientist whose opinions on the health risks of cell phones were widely sought after by organizations such as the International Commission on Non-Ionizing Radiation Protection (ICNIRP), the World Health Organization (WHO), the Swedish Radiation Protection Authority, and the European Union.

In 2011, Ahlbom was set to serve on an expert panel organized by WHO's International Agency for Research on Cancer (IARC). That was until an investigation by Swedish journalist Mona Nilsson discovered that Ahlbom, along with his brother Gunnar, owned and served on the board of a lobbying firm servicing multiple firms in the telecommunications industry. Ahlbom's clients included global telecommunications giant Ericsson (whose networks, at the time in 2011, handled 40% of all mobile phone calls made in the world)[74] and Swedish mobile phone operator TeliaSonera. Though the IARC requires invited experts like Ahlbom to disclose any such ties, Ahlbom did not.[75] The IARC effectively disinvited Ahlbom's participation in the panel. Shortly thereafter, Ahlbom resigned his position with the Swedish Radiation Protection Authority just weeks after that group opened an investigation into Ahlbom's potential conflicts of interest.

Precisely how such undisclosed conflicts of interest may have influenced policy is difficult to say. Though, as you've seen, Ahlbom was a respected expert in a position of significant authority when it came to establishing regulations for mobile devices around the world. From his own words we can glean what type of influence he has had. In a 2011 interview, Ahlbom was quoted (translated from the original Swedish) as saying "the probability of mobile phone radiation causing brain tumors is low . . . We are now pretty sure that there is no relation [between mobile phone use and brain tumors], at least after 10–12 years of use . . . Furthermore, there are areas that have not been studied, for example mobile phone use among children and youth. There is, however, no reason to believe that there are any risks there either."[76] (This last comment is particularly egregious given the increased risks of EMF exposure for children, as described in chapter 13.)

Shortly after Ahlbom's departure from the IARC (a division of WHO), WHO voted to classify MW radiation as a class 2B carcinogen—meaning that there is evidence, but not definitive proof, linking microwave radiation and cancer. Sadly, Ahlbom is just one example of the extreme conflicts of interest that can cloud and distort scientific research.

In 2002, the Swedish Radiation Protection Authority hired Drs. John Boice Jr. and Joseph McLaughlin, from the privately held firm International Epidemiology Institute (IEI), to review the published epidemiology on cell phone use and cancer. Boice and McLaughlin concluded that there was no consistent evidence for any increased risk of brain or salivary gland cancer resulting from cell phone use. These two men, however, failed to mention that they were the authors of some of the studies that they had reviewed—studies that demonstrate no correlation between mobile phone radiation and specific types of tumors. Boice and McLaughlin also failed to disclose that their employer, IEI, was involved on behalf of Motorola in a case involving cellular phones and brain tumors (*Newman v. Motorola Inc.*, 2002).[77]

The wireless industry not only has significant resources to fund its message, but as you've seen, the underlying science is extraordinarily complex and easy to obfuscate. This makes it very difficult for most people (including judges and juries) to interpret the information. The creation of misleading science to combat public perception through the media has been one front in industry's war on science. Another key fight has been the battle to prevent any government regulation that might reduce profitability. On this front, as well, the wireless industry has been phenomenally successful.

FIGHTING REGULATION

As noted earlier in this chapter, the EPA's research into the health effects of EMF and cell phone radiation was defunded in the 1990s. This is just one example of how the wireless industry has been able to influence the direction of federal government regulation. Between 1998 (when the wireless industry was first required to disclose its expenditures on lobbying efforts) through 2005,

> eight of the country's largest and most powerful media and telecommunications companies, their corporate parents, and three of their trade groups, have spent more than $400 million on political contributions and lobbying in Washington, according to a Common Cause analysis of federal records. Verizon Communications, SBC Communications Inc., AOL Time Warner, General Electric Co./NBC, News Corp./Fox, Viacom Inc./CBS, Comcast Corp., Walt Disney Co./ABC, and the National Association of Broadcasters (NAB), the National Cable Telecommunications Association, and the United States Telecom Association together gave nearly $45 million in federal political donations since 1997. These eight companies and three trade associations also spent more than $358 million on lobbying in Washington, since 1998, when lobbying expenditures were first required to be disclosed.[78]

One of the largest victories for the wireless industry in the fight against regulations was the 1996 Telecommunications Act (TCA). Though this occurred before lobbying expenditures were required to be disclosed, it is reported that the wireless industry spent $50 million influencing the outcome.[79] Included as section 704 was language specifically barring any restrictions on placement of cell towers due to health concerns.

As a result, many of the battles today over EMF regulation occur at the state and local level. In 2011, the California Council on Science and Technology (CCST) invited experts to comment on its *Smart Meter Report* regarding the increasingly common new generation of power meters that report power usage details back to the utility using wireless RF communication. (The Federal Communications Commission rates smart meters, like cell phones, as a safe technology.) The CCST report concluded that smart meters were not a health risk.

Interestingly, the final report (on which media outlets such as AP were basing their reporting) failed to include several comments, including many from experts that CCST had explicitly invited to comment. One omitted comment was the following one from Dr. De-Kun Li from the Kaiser Foundation Research Institute:

> The bottom line is that the safety level for RF exposure related to non-thermal effect is unknown at present and whoever claims that their device is safe regarding non-thermal effect is either ignorant or misleading.[80]

Another, more specific, comment that was not included is by Dr. Daniel Hirsch from the University of California at Santa Cruz:

> [The report's] estimates appear incorrect in a number of regards. When two of the most central errors are corrected . . . the cumulative whole body exposure from a Smart Meter at 3 feet appears to be approximately two orders of magnitude higher than that of a cell phone, rather than two orders of magnitude lower.[81]

It's not clear why these comments, which CCST solicited, were not included in the report. Whatever the reason, the result was clear: invited comments from experts challenging the conclusions that smart meters are safe were omitted from a state-level report that influences regulation of the technology.

Another example from California is from 2010, when San Francisco became the first city in the nation to mandate that all stores include SAR (specific absorption rate) ratings alongside pricing for all cell phones. It seems a reasonable enough policy measure, providing consumers with information on EMF radiation levels generated by different devices. After all, cell phone makers have to disclose this anyway; this law just enforces a more prominent display of the same information. Unfortunately, Mayor Gavin Newsom explains, this is not what happened. Instead, "lobbyists from Washington made it clear that they would invoke 'the nuclear option' and come down 'like a ton of bricks.'"[82] As one example, Newsom explains that the Marriott hotel chain sent him a letter reading:

> CTIA – The Wireless Association, which is scheduled to hold a major convention here in October 2010, has already contacted us about canceling their event if the legislation moves forward. They also have told us that they are in contact with Apple, Cisco, Oracle and others who are heavily involved in the industry, as you know, about not holding future events in your city for the same reason.[83]

Immediately following passage of the bill, CTIA announced that it would pull its annual convention, with 68,000 attendees and an estimated $80 million in business, away from San Francisco. On the experience, Newsom reflects:

> Since our bill is relatively benign, it begs the question, why did they work so hard and spend so much money to kill it? I've become more fearful, not less, because of their reaction. It's like BP. Shouldn't they be doing whatever it takes to protect their global shareholders?[84]

CONCLUSION

While there are many dedicated scientists who are searching for the truth in regard to the dangers of EMF, they are an endangered species in the United States. The government no longer funds this research. The wireless industry funds studies that produce results in line with their interests, and attacks and defunds those scientists who produce results contrary to their interests. As Carlo explains, "the industry strategy has been to fund low-risk studies that will assure a positive result—and then use it to convince the news media and the public that it is proof that cell phones are safe."[85] Jerry Phillips, who has seen this play out firsthand, reiterates that the wireless industry is "not interested in solving scientific puzzles, they're interested in making money."[86]

And making money is definitely something at which the wireless industry has excelled. Estimates are that the wireless industry as a whole netted $19 billion in profits in the first quarter of 2012—reflecting a 20% increase from 2011.[87] The industry is so profitable that it can fund a significant amount of research—enough to obscure the value of those studies that do demonstrate health effects from cell phone use. Industry influence on EMF science comes at a time when there is no US-government research program at all on EMF safety—while use of cell phones, WiFi networks, and new wireless technologies like smart meters are all dramatically increasing.

While the bulk of examples in this chapter are of the corporate role in EMF science, much the same is found with military research sponsorship. Dr. Allan Frey reports that, following his groundbreaking results demonstrating blood-brain barrier damage from EMF exposure (discussed in chapter 6), his sponsor, the Office of Naval Research (ONR), instructed him not to publish further or he would lose his funding.[88] Similarly, Dr. Milton Zaret's early pioneering work into microwave cataracts (also discussed in chapter 6) led to the cessation of all military funding for his research (which is why neither he nor anyone else has had the opportunity to attempt replication of the results).

As Chris Mooney writes in the *Prospect*, "the sabotage of science is now a routine part of American politics . . . It happens virtually every time the government even dreams of regulating a substance."[89] Indeed, as we'll see in the next chapter, the sabotage of EMF science (and the resulting delay of government regulatory action) is highly reminiscent of the history of another controversial product—cigarettes.

Chapter 9

DOUBT, FROM TOBACCO TO INTERPHONE

In the novel and film *Thank You for Smoking*, character Nick Naylor is a phenomenally successful lobbyist for the tobacco industry. An unpredictable set of circumstances, however, ruins his career. All seems lost until he has an epiphany—he realizes that he can transfer his skills to a new set of clients. At the close of the film, we see a rejuvenated Naylor advising a team of anonymous executives to repeat the following mantra to the media and the public: "Although we are constantly exploring the subject, currently there is no direct evidence that links cell phone usage to brain cancer."

And so we learn that Nick Naylor has begun a promising new career as a wireless-industry lobbyist, applying the same skills—and quite literally the same message—that he honed in his years serving tobacco companies.

CELL PHONES AND CIGARETTES

The comparison between cell phones and cigarettes has been made elsewhere. When Maureen Dowd wrote her 2010 New York Times op-ed column entitled "Are Cells the New Cigarettes ?," she helped raise awareness of the important public health risks associated with wireless communication by directly linking mobile phones to tobacco: "Just as parents now tell their kids that, believe it or not, there was a time when nobody knew that cigarettes and tanning were bad for you, those kids may grow up to tell their kids that, believe it or not, there was a time when nobody knew how dangerous it was to hold your phone right next to your head and chat away for hours."[1]

Similarly, in 2008 FOX News announced, "Study: Cell Phones

Could Be More Dangerous Than Cigarettes,"[2] and three years before that, CNET's Molly Wood wrote an article entitled "The Cell Phone Industry: Big Tobacco 2.0?"[3]

These cultural references point to a shared health risk from the use of these products. However, the most notable similarity between the wireless and tobacco industries involves the ways in which these companies have responded to scientific research that challenges their profit models.

DOUBT IS OUR PRODUCT

You'll note that Nick Naylor does not advise the executives to claim that cell phone usage is benign. Instead, he suggests challenging the existence of any *proven* link. In other words, don't claim "cell phones are safe." Instead, claim "there's no proof cell phones are dangerous." And this is frequently what we hear. Though some executives go so far as to claim that cell phones are nontoxic, you'll note that, more typically, they state that there is a lack of conclusive proof that non-ionizing EMF is carcinogenic.

In other words, the wireless industry has adopted a strategy of manufacturing doubt about the potential negative health effects of their product. Sadly, this strategy has a proven track record. The tobacco industry used it for decades to fend off regulation and negative public perceptions. As one executive from Brown Williamson (BW), a large tobacco company, infamously wrote in a 1969 memo eventually leaked to the public: "Doubt is our product, since it is the best means of competing with the 'body of fact' [linking smoking with disease] that exists in the mind of the general public. It is also the means of establishing a controversy."[4]

TOBACCO

Industry's role in scientific research (such as that described in the previous chapter) is by no means new. A range of controversial industries have been involved in conducting studies to support the safety of their products. But no industry has managed this type of public relations dilemma more effectively than the tobacco industry. Between the 1920s and the 1950s, tobacco companies used deceptive and often blatantly false claims in an effort to reassure the public that their products were safe. For years, consumers had been exposed to advertisements showing the value of smoking. And backing up the claims were pictures of doctors advocating the benefits not simply of smoking, but of smoking particular brands. What emerged were highly successful, evocative advertising campaigns that strategically used doctors and celebrities to endorse cigarettes. *More doctors smoke Camels than any other cigarette!* was the famous catch phrase for a Camels advertising campaign that began in 1946 and ran for eight years in both magazines and on the radio.

Over the years (since the passage of time is required for detrimental health effects to emerge), serious concerns began to emerge about the dangers of smoking. Those led to increasing calls for the regulation. Regardless of the validity of the concerns, the tobacco industry mounted major campaigns to prevent any and all restrictions.

A key part of the tobacco industry's efforts involved hiring scientists to conduct seemingly sophisticated studies ostensibly aimed at determining whether smoking was, in fact, dangerous. Over and over again, their results showed that no clear determination could be made. The studies were summarized in 2008 in *Doubt Is Their Product* by scientist and former government regulator David Michaels. Michaels served as Assistant Secretary of Energy for Environment, Safety, and Health during the Clinton administration. As Michaels so astutely points out, the industry sought to create doubt about the health charge without actually denying it. It was a highly effective strategy to fend off any regulatory action or corporate responsibility for the fatal impact of their actions. The industry did not conduct research to find out the facts; it carried out research to create enough doubt so as to undermine any challengers' claims and thereby block any action. "No industry has employed the strategy of promoting doubt and uncertainty more effectively, for a longer period, and with more serious consequences," writes Michaels.[5]

"This era of over-the-top hucksterism went on for decades, and it was all blatantly false," said Dr. Robert J. Jackler of the Stanford School of Medicine.[6] Jackler produced a fascinating, retrospective look at the advertising crusade that defined the tobacco industry's cunning tactics in an exhibit entitled *Not a Cough in a Carload: Images Used by Tobacco Companies to Hide the Hazards of Smoking*.

Meanwhile, the scientific evidence continued to build, leading to the first major study that causally linked smoking with lung cancer. Published in 1950 by American scientists Ernst L. Wynder and Evarts A. Graham, the report indicated that 96.5% of lung-cancer patients are moderate to heavy smokers.[7] Ironically, the *Journal of the American Medical Association* (JAMA), which published this study, simultaneously ran cigarette advertisements. Wynder followed up with another landmark study showing that painting cigarette tar on the backs of mice created tumors in 44% of the mice within a year of such exposure.[8]

Even so, the advertisements persisted. One for Chesterfield cigarettes in 1952 claims, "no adverse effects on nose, throat and sinuses" after medical specialists observed smokers for 10 months. That same year, a significant turning point came when an influential article in *Reader's Digest* titled "Cancer by Carton" detailed the dangers of cigarettes for the mainstream public. Within a year, cigarette sales fell for the first time in more than two decades.

THERE IS NO ESTABLISHED LINK

The word had finally reached the people, and the tobacco industry had to scramble. They decided that the best way to refute science was with science. So, in 1954, the tobacco industry launched the "Sound Science" campaign by creating the Tobacco Industry Research Committee (TIRC), which later became the Council for Tobacco Research (CTR). It purported to fund independent scientific research, asserting that public health was "paramount to every other consideration in our business."[9]

TIRC launched a multifaceted, multinational strategy to mislead consumers about the established dangers associated with smoking cigarettes. In a true blitz campaign, it ran a full-page promotion in more than 400 newspapers aimed at an estimated 43 million Americans titled "A Frank Statement to Cigarette Smokers." Its opening words read:

> RECENT REPORTS on experiments with mice have given wide publicity to a theory that cigarette smoking is in some way linked with lung cancer in human beings.
>
> Although conducted by doctors of professional standing, these experiments are not regarded as conclusive in the field of cancer research. However, we do not believe that any serious medical research, even though its results are inconclusive should be disregarded or lightly dismissed.
>
> At the same time, we feel it is in the public interest to call attention to the fact that eminent doctors and research scientists have publicly questioned the claimed significance of these experiments.[10]

During the 1950s, tobacco companies greatly increased their advertising budgets from $76 million in 1953 to $122 million in 1957. The TIRC spent another $948,151 in 1954 alone and referred to it as the "1954 emergency."

This continues even today. As just one recent example, in 2002 Dr. Ragnar Rylander, Professor of Environmental Health at Gothenburg University, was revealed to have received significant sums in research grants and consulting fees from Philip Morris *over a 30-year period.* Without disclosing this financial relationship to his employer, or in any of the multiple papers that he published, he presented data minimizing and denying negative health effects of tobacco and secondhand smoke. Rylander repeatedly denied these accusations until his contract was uncovered in the Philip Morris archives.[11] A Swiss court in Geneva subsequently found that Rylander failed to fulfill his "moral obligation" to disclose these financial ties,[12] declaring in their final ruling that

"Geneva has indeed been the centre of an unprecedented scientific fraud in so far as Ragnar Rylander, acting in his capacity of associate professor at the University, took advantage of its influence and reputation, not hesitating to put science at the service of money and not heeding the mission entrusted to this public institution."[13]

Eventually, the overwhelming body of scientific evidence led to the 1964 US Surgeon General's report citing health risks related to smoking. The following year, the US Congress passed the Federal Cigarette Labeling and Advertising Act, requiring a surgeon general's warning on cigarette packs. But instead of putting the kibosh on Big Tobacco's activities, the FDA warning label conveniently provided them with a legal loophole that removed any corporate responsibility. And so, they continued selling a product that kills, promoting the "safe cigarette" in the 1970s and then battling the regulation of secondhand smoke in the 1980s. Meanwhile, the death toll continued to rise.

WHISTLE-BLOWER

The breaking point finally came when Jeffrey Wigand, a BW tobacco executive, put himself on the line as a whistle-blower. He found himself compelled to speak out after watching the 1994 US congressional hearing where eight Big Tobacco executives testified under oath that "nicotine was not addictive." As the head of research and development for BW, Wigand knew otherwise. As depicted in the dramatic film *The Insider*, Wigand's insider testimony revealed that Big Tobacco was consciously deceiving the public and ultimately led to the multibillion dollar settlement in 1997 that required tobacco companies to pay out $368 billion in health-care costs due to smoking-related illnesses.

Wigand's testimony has been further supported by millions of internal documents that have since come to light—many of which were publicly released by Stanton Glantz, a professor of medicine at the University of California at San Francisco, in his book *The Cigarette Papers*. A long-time critic of the tobacco industry, Glantz received an anonymous package in 1994 delivered to his office at the University of California at San Francisco. It contained more than 4,000 pages of internal tobacco-industry documents sent by a secret source named "Mr. Butts." A thorough analysis of the documents revealed in detail the strategies behind Big Tobacco's deceitful practices and exposed the fact that they knew about the health dangers all along.

Today, approximately 80 years after German scientists produced the first data suggesting a possible link between tobacco smoking and lung cancer,[14] the health risks of smoking are clearly understood by the public and the media, tobacco sales and smoking in public establishments are more tightly regulated, and some cities around the United States have even gone so far as to institute outright bans on smoking anywhere within their borders (other than private homes). But it took a long time. Tobacco's strategy of creating doubt successfully defended significant profits for decades as millions died from lung cancer.

Is the wireless industry following the same playbook, perfected by the tobacco industry? To help answer this question, it is instructive to look at the recently concluded Interphone study.

INTERPHONE

The large-scale Interphone study was initiated in 2000 and formally concluded in February 2012 (with some results having been released earlier), and its results received a great deal of media coverage. Interphone was created by the International Agency for Research on Cancer (IARC), a division of the World Health Organization (WHO). The actual research was conducted at 25 separate research institutions, across 13 countries.[15] The IARC coordinated these participating institutions and their funding; the IARC's Interphone International Study Group (IISG), with 21 scientists led by Dr. Elisabeth Cardis, administered the progress of the study and how the data was analyzed, interpreted, and published.

Interphone's goal was clear and simple: to evaluate whether any link could be established between cell phone usage and occurrences of four types of cancer in human tissue that is most exposed to cell phone radiation: glioma and meningioma (tumors of the brain), cancer of the parotid gland (a type of salivary gland), and schwannoma (tumors of the acoustic nerve).[16]

In general, with epidemiological studies (studies of health issues in a population), the greater the number of subjects, the more accurate the findings are apt to be—it is not advisable to draw conclusions when the number of subjects is relatively small. With Interphone, this was not a problem. Data was collected from large populations in all 13 participating countries. Indeed, the studies that took place in Denmark, Finland, Israel, Norway, and Sweden encompassed almost the entire population of each nation.[17] Interphone produced 3 results papers,[18] 4 validation studies,[19] and 36 individual studies from the individual participating research institutions.[20] All told, Interphone represents the largest case-control study of mobile phone use and these types of cancer in subjects with at least a decade of reported exposure.

The study found that cell phone usage (including regular usage for a decade or longer) is not linked to increased risk of brain tumors (specifically, glioma or meningioma)—though there may be some relationship between heavy cell phone usage (the top 10% of users) and up to a 40% greater risk for occurrence of gliomas on the same side of the head as cell phone use. The overall results of the study are summarized in the final IARC report:

> Overall, no increase in risk of glioma or meningioma was observed with use of mobile phones. There were suggestions of an increased risk of glioma at the highest exposure levels, but biases and error prevent a causal interpretation. The possible effects of long-term heavy use of mobile phones require further investigation.[21]

INTERPHONE FUNDING

Of course, running such a large study costs a significant amount of money—€19.2 million (roughly $25.5 million in 2010 dollars) according to IARC's 2010 report. Approximately 29% of this funding (roughly $7.4 million in 2010) was provided by wireless-industry sources and the remainder from European public institutions (the US government did not participate in Interphone). The Mobile Manufacturers' Forum (MMF) and the GSM Association donated €3.5 million of funding (roughly 2010 US $4.6 million)—though the Union for International Cancer Control (UICC) established a funding firewall to help ensure scientific independence.[22] The IARC states that the organizations that funded Interphone did not have access to any of the results prior to publication (though they, along with other specific groups, could review the articles seven days prior to their actual publication).[23] However, the fact that the wireless and cell phone industries paid for nearly half of the study's cost by definition calls into question the study's independence.

The existence of controls put in place by the IARC to separate the scientists from the pressure of funding was admirable. However, one could have more faith in the integrity of the process if the IARC would make public its conflict-of-interest statements from the Interphone project—something that the IARC and WHO have consistently refused to do. In the absence of such disclosures, observers are left with little confidence that there are no Rylanders or Ahlboms—or, indeed, Anders Ahlbom himself, who was still in good standing for the early part of Interphone's history—and possibly influencing or skewing the research.

INTERPHONE'S DESIGN FLAWS

Given the massive scope of the study, Interphone would seem to go a long way toward settling the issue. Except, it is important to note the following part of the above-cited conclusion: "biases and error prevent a causal interpretation." How can it be that the largest such study, spanning hundreds of thousands of participants, across 13 countries, costing tens of millions of dollars could produce results that are so riddled with "biases and error" as to prevent interpretation?

The biases and errors to which the IARC refers are found largely in the design of the study—the type of data Interphone was structured to collect and how it was collected—factors that should have been avoided in the planning stage. One of the most significant design flaws was the researchers' reliance on people's memories. The wireless networks would not permit researcher access to actual cell phone records, so researchers resorted to asking individual subjects to recall their cell phone usage over time. Recalling how much time you spent on your phone last week or last month is difficult enough—asking participants to accurately estimate how much they used their phones 10 years ago is a pretty unreliable form of data collection.

Reliance on people's memories in scientific studies skews the results with something called *recall bias*, generally yielding highly unreliable data. Indeed, one of the Interphone studies outfitted subjects with special equipment to track their cell phone usage, generating a detailed and accurate log. "When this log was compared with the 'recalled' usage, there were wide and random variations: some users underreported, while others overreported use."[24] And, of course, any study that relies on an individual's memory, also implicitly relies on that individual being alive; this eliminates any potential subjects who may have already died from the cancers being investigated (leading to an underestimation of risk in the results).

Overall, Interphone had significant difficulty recruiting respondents. The refusal rate was 41%—a rate that many statisticians believe taint any results garnered from the respondents who do accept.[25] This creates what researchers call *selection bias*, or a misrepresentation of actual populations in the study, leading to unreliable results. Another way in which the study's design generated selection bias was in the location of respondents. As I mentioned, in five countries Interphone collected nationwide data. However, in seven others,[26] the data were based on responses from subjects primarily residing in urban centers.[27] Those who live in urban areas are likely to be closer to a cell tower than those who live in a rural setting. The farther one is from a cell tower, the more power one's cell phone must generate in order to communicate with the tower. Thus, the elimination of rural subjects excluded those who, in general, are exposed to the most powerful EMF radiation from their phones.

Another significant selection bias in the study was the omission of young subjects. Children and adults up to 30 (along with those 60 and older) were excluded from Interphone, again, by design. However, children (who continue to grow and undergo a higher rate of cellular division and reproduction than adults) are more susceptible to developing cancer from exposure to carcinogens. (And, as we know, children and twentysomethings do use cell phones.) Excluding this higher-risk population from the study would lead to an undercalculation of risk in the results.

Another design flaw regards latency time, or the length of time it takes the cancer to develop. Most cancers have latency periods of over 10 years; brain tumors are believed to take up to 25 years to form. Thus, analyzing a period of 10 years would likely not provide sufficient time for brain tumors or other cancers to become symptomatic, leading to an undercalculation of risk in the results. Jack Siemiatycki, the Canada Chair in Environmental Epidemiology at the University of Montreal, explains that "if it turns out that cellphones cause brain cancer, but in a 15- to 20-year time period before the tumours manifest themselves clinically, we [Interphone] would not have been able to pick this up."[28]

Further, while Interphone was designed to survey 10 years' worth of cell phone use (an admirable goal, given the shorter duration of many other studies), cell phone use was not widespread in 1990 (10 years prior to the initiation of the study)—and even by 1994 (10 years before the end of Interphone's data-collection period), cell phone use was still quite small by today's standards. While study subjects may have been active cell phone users in 2000, they were likely not in 1990 or 1994, and thus most respondents would not have had sufficient duration for the cancers to become symptomatic.

For example, of the nine Interphone studies relating to brain tumors, only 0.61% of subjects who reported cancers and 10% of controls had used a cell phone for 10 years or more; cell phone use of greater than five years was reported among 18% of reported cancers and 21% of controls.[29] In other words, the results included only a very small number of long-term cell phone users—so small that it is impossible to draw any conclusions regarding the possible link between cell phone use and brain tumors.

Interphone's definition of exposure in itself is one that would lead to an underassessment of risk. By design, a regular user was defined as anyone who had an average of at least one phone call per week for 6 or more months—this definition classified individuals with very low exposures to cell phone radiation in the group of "regular" users, thus underassessing risk. Further, Interphone's subject matter was restricted to cell phones; individuals who used cordless phones at home did not count as having been exposed. Cordless phones, like cell phones, emit RF/MW radiation. Still, even though these individuals were exposed to the same type of EMF radiation, in the same location of the body, Interphone classified these individuals as unexposed, skewing the results with increased rates of negative health outcomes from EMF exposure among those designated "unexposed."

And, of course, by design, Interphone examined only four types of cancer. However, as we've seen in our investigation, reputable scientific research has linked RF and MW exposure to many different types of cancers (such as leukemia, melanoma, and lymphoma), as well as other negative health outcomes in humans, such as Alzheimer's. Restricting the investigation to four types of cancers is necessarily going to underrepresent any risk from EMF exposure.

DELAYS AND CONFUSION

Interphone had many design flaws that would preclude the possibility of reaching useful conclusions. Not surprisingly, this led to feuding among the participating scientists over the interpretation of data,[30] which generated significant delays in the release of the results—delays that the European Parliament eventually condemned as "deplorable."[31] The data-collection phase of Interphone ended in 2004, and the results were scheduled for publication in 2006.[32] Some studies were published separately, but the official, final Interphone report was not published and made available to the scientific community until February 2012 (though, as can occur in scientific studies, it is dated October 2011).[33] IARC formally shuttered Interphone a few days later.[34]

Overall, Interphone concludes that heavy cell phone usage could be linked to brain tumors (though "biases and error prevent causal interpretation"). Confusingly, the report also indicates that the data reveals a reduced risk of brain tumors stemming from cell phone use—a baffling conclusion that has led "most epidemiologists, including the authors of Interphone, to consider that "the results point to a systemic flaw in the trial."[35]

Interphone also concludes that no link exists between cell phone use and acoustic neuromas (cancer of the auditory nerve)—though again, one Interphone study with data from five participating countries reported an 80% greater risk on the side of the head where the cell phone is held after ten years of use.[36] In the end, the IISG refused to pool and analyze the data on tumors of the partoid gland—leaving individual countries to report individually (such as the work by Dr. Sadetzki in Israel, discussed in chapter 5).

In short, each of the findings was ambiguous, leaving room for individual interpretation of results. Interphone itself explained that no real conclusions could be drawn. Exacerbating matters, the IARC has refused to release the actual data collected under the Interphone project (though some individual researchers have published subsets of the data). Dr. Lennart Hardell's highly regarded research on brain tumor risk from cell phone use, discussed earlier in this book, is challenged by the Interphone reports, and he has asked that the data be made available. However, despite repeated pleas from him and many other investigators, the only study-wide information provided are the published reports containing the ambiguous analyses.[37] As a result, despite the amount of time, money, effort, and data that Interphone represents, the study and its conclusions are almost entirely devoid of any scientific value.

MEDIA COVERAGE OF INTERPHONE'S RESULTS

While Interphone produced data that was, at a minimum, questionable due to the amount of "biases and errors," it represented such a large effort that the results were widely reported in the global media. Not surprisingly, the confusing and often contradictory analyses of the findings meant that the media coverage was similarly confusing and contradictory, doing little to help the public understand this complex issue. CNN led with the headline "Study Fails to End Debate on Cancer, Cell Phone Link," explaining that "long-awaited data from an international study have shown no evidence of increased risk of brain tumors associated with mobile phones, except in people who have the most exposure. But design flaws of the Interphone study, which is partly industry funded, suggest that the latest results cannot be taken to mean that cell phones and brain cancer are unrelated, critics say."[38]

That's a fair enough reading of the IARC analysis, but it does little to help explain the Interphone results to the average reader. What's more, the results were sufficiently ambiguous that Interphone researcher Daniel Krewski of the University of Ottawa in Ontario could explain, in the very same CNN article, that the study was "scientifically sound" and produced "reassuring" results. "It tells us that we don't have an epidemic of brain cancer on our hands associated with mobile phones."[39]

The *New York Times* covered the 2010 results in an article entitled "Do Cellphones Cause Brain Cancer?" The author explains that "trials like Interphone are undertaken in the hope that they cleanse the field of doubts. In fact, Interphone achieved just the opposite effect: it ignited even more puzzling questions. Over all, the study found little evidence for an association between brain tumors and cellphones."[40]

As the *Wall Street Journal* sarcastically stated regarding the study's results, "using a cellphone seems to protect against two types of brain tumors. Even the researchers didn't quite believe it."[41] The journal *Nature* reported on Interphone in an article entitled "No Link Found between Mobile Phones and Cancer"; though, despite the title, in the article the author admits that "unfortunately, the results from this study are not entirely straightforward . . . Even the researchers involved in the trial do not all agree on the meaning of their work."[42]

Demonstrating just how confusing the Interphone study could be, FOX News reported that "an increased risk of brain cancer is not established from the data from Interphone,"[43] while in a separate story it reported that "the WHO's Interphone investigation's results showed, 'a significantly increased risk' of some brain tumors 'related to use of mobile phones for a period of ten years or more.'"[44] Around the world, the picture was much the same.

Unfortunately, public health groups did little to clarify or interpret the results, echoing the same message as reported in the media. The American Cancer Society explained that Interphone's results "do not establish a definitive link between cell phone use and cancer, but they don't rule one out, either . . . [It] may have raised more questions than it answered."[45] The World Health Organization claimed that "to date, no adverse health effects have been established for mobile phone use . . . an increased risk of brain tumors is not established from Interphone data," though they proceeded to call for more studies into long-term exposure.[46] The Independent Advisory Group on Non-Ionizing Radiation (AGNIR) of the British Health Protection Agency concluded that Interphone "provides no clear, or even strongly suggestive, evidence of a hazard," adding that this "is consistent with the findings of most other epidemiological studies that have examined the relation of brain tumours to use of mobile phones."[47] They concluded this even though the UK Health Protection Agency itself reported that "biases and errors" within the study have restricted conclusions that can be drawn.[48]

Some of the media summaries were less accurate, omitting mention of the possible increased risk of brain tumors for heavy cell phone users, demonstrated by Interphone. Then internationally respected Professor Anders Ahlbom (prior to his outing as an industry-paid stooge discussed in the previous chapter)—who had repeatedly served as an expert on cell phone radiation for WHO and had helped establish EU cell phone radiation safety standards—explained that "Interphone shows the same results as all other research studies so far, namely that there is nothing to be worried about."[49] Similarly, the Italian National Institute of Health concluded that "overall, the study do [*sic*] not report any increase of risk for brain tumors associated with mobile phone use, not even among the long-term user (more than 10 years)."[50] Similarly, the US National Cancer Institute reported that "cell phone users have no increased risk of two of the most common forms of brain cancer . . . There was no evidence of risk with progressively increasing number of calls, longer call time, or time since the start of the use of cell phones."[51]

THE IMPACT OF INTERPHONE

From a review of these and similar articles and reports, one could reasonably conclude that Interphone was both meaningful and unreliable, that the results demonstrate no cause for alarm. "This study did not confirm or dismiss the possible association between cell phones and brain tumors. That's the bottom line," summarizes Interphone researcher Dr. Siegal Sadetzki.[52]

Interphone represents the largest and most ambitious research endeavor into the epidemiology of the negative health effects of exposure to EM radiation—particularly, in this case, RF and MW radiation emitted by cell phones. If the studies had been well designed, the results could have been invaluable. Instead, what we see are ambiguous and vague conclusions drawn from fundamentally flawed data, with the discussion in the media being no better informed as a result.

Even worse, as Professor Jorn Olsen at the School of Public Health of Aarhus University in Denmark explains, not only should the Interphone funding "have been used better by setting up a large-scale cohort study that could address other potential endpoints besides cancer,"[53] but "the Interphone Study dried up available resources for funding and made the public and funding agencies immune to the epidemiological results."[54] There is not enough money spent on this type of research to begin with; Interphone consumed a significant portion of it for the better part of a decade.

It's not necessarily the case that Interphone was designed to produce faulty data; however, if one wanted to create a study that generated largely useless results, the design of Interphone would be an effective means of doing so. Similarly, it may have been unintentional that the study extended eight years longer than planned and consumed much of the funds available for this kind of research, but this reduced the number of competent studies that could have been done to answer the questions of health risk. What a waste of time and money!

IS EMF THE NEW TOBACCO?

Similar to the tobacco industry's infamous "doubt is our product" memo, another leaked document outlined Brown and Williamson's objectives at the time:

> Objective No. 1: To set aside in the minds of millions the false conviction that cigarette smoking causes lung cancer and other diseases; a conviction based on fanatical assumptions, fallacious rumors, unsupported claims and the unscientific statements and conjectures of publicity-seeking opportunists.
>
> Objective No. 2: To lift the cigarette from the cancer identification as quickly as possible and restore it to its proper place of dignity and acceptance in the minds of men and women in the marketplace of American free enterprise.
>
> Objective No. 3: To expose the incredible, unprecedented and nefarious attack against the cigarette, constituting the greatest libel and slander ever perpetrated against any product in the history of free enterprise . . .
>
> Objective No. 4: To unveil the insidious and developing pattern of attack against the American free enterprise system, a sinister formula that is slowly eroding American business with the cigarette obviously selected as one of the trial targets.[55]

With the exception of the fourth objective (defenders of the wireless industry tend to imply their opponents are Luddites rather than communists), the others sound very familiar. All you have to do is remove the word "lung," replace "cigarette" with "cell phone," and "smoking" with "usage."

> Objective No. 1: To set aside in the minds of millions the false conviction that cell phone usage causes cancer and

> other diseases; a conviction based on fanatical assumptions, fallacious rumors, unsupported claims and the unscientific statements and conjectures of publicity-seeking opportunists.
>
> Objective No. 2: To lift the cell phone from the cancer identification as quickly as possible and restore it to its proper place of dignity and acceptance in the minds of men and women in the marketplace of American free enterprise.
>
> Objective No. 3: To expose the incredible, unprecedented and nefarious attack against the cell phone, constituting the greatest libel and slander ever perpetrated against any product in the history of free enterprise . . .

The analogy of cell phones to tobacco naturally begs the question, does the wireless industry know more than it is letting on? As a scientist, I am struck by the volume of scientific evidence that has been conveniently overlooked or dismissed when it comes to the EMF "debate" and the apparent "controversy" when it comes to protecting public health. When looking at the techniques and arguments used by the wireless industry, I have no doubt that doubt is their product.

Some insider documents have already begun to surface that indicate that the cell phone industry indeed knows more about the EMF health dangers than they are letting on. Two in particular have been released under the Freedom of Information Act to *Microwave News*. The first was the "war-games" memo, discussed in the previous chapter, revealing Motorola's intent to discredit Dr. Henry Lai and his work.

The other was written in 1993, the same year that the Reynard lawsuit hit national headlines and jump-started the industry's $25-million WTR initiative. This internal memo at the Food and Drug Administration (FDA) noted that the data "strongly suggest" that microwaves can "accelerate the development of cancer." It went on to give supporting details: Drs. Mays Swicord and Larry Cress of FDA's Center for Devices and Radiological Health (CDRH) in Rockville, Maryland, wrote, "Of approximately eight chronic animal experiments known to us, five resulted in increased numbers of malignancies, accelerated progression of tumors, or both." (This is the same Mays Swicord discussed in the prior chapter, who later became director of research at Motorola and advocated ending research on biological effects of EMF.) However, the FDA played down these findings in the public statements at the time and subsequently abandoned the oversight of the CTIA's research program, leaving it in the sole care of the wireless industry.

SCIENTIFIC UNCERTAINTY

What we see happening today with EMF science (and over the past 90 years with tobacco science), has happened repeatedly, across a variety of industries involving global warming, asbestos, lead, plastics, and other toxic materials. Companies have regularly skewed the scientific literature, manufactured and magnified scientific uncertainty, and influenced policy decisions to keep the public confused.

History shows us that it can take 30 to 100 years between the first early warning signs and the regulatory action taken to protect public health (see chart on page 154). The scientific evidence for EMF, however, has finally reached a tipping point. It is calling us to stop arguing the science, and to move into acceptance that EMFs are indeed hazardous.

Unfortunately, the success of the wireless industry at manufacturing doubt has significantly slowed the progress in the establishment of biologically based safety standards to protect consumers. As we will see in the next chapter, safety standards and regulatory frameworks around the world are based on flawed and outdated assumptions about the health risks associated with EMF exposure.

Late Lessons Chapter	First Early Warning	Date of Effective Risk-Reduction Action	Years of Inaction
Fisheries: taking stock, overfishing	1376	1995–2008	hundreds
Radiation: early warnings, late effects	1896	1961–96 (UK etc., then EU laws	65–100
Benzene: Occupational setting	1897	1978 (benzene withdrawn from most consume	81
Asbestos: from "magic" to malevolent material	1898	1999, EU ban by 2005	101–7
PCBs and the Precautionary Principle	1899	1970–80s (EU and US restrictions; phase out by 2010	c. 100
Halocarbons, ozone layer, and Precautionary Principle	1974	1887–2010 (global ban on CFCs and other Ozone depleters)	10–30
DES: long-term consequences of prenatal exposure	1938	1971–85 (US, EU, global ban)	33–47
Antimicrobials as growth promoters	1969	1999 (EU ban)	30
SO2: from protection of human lungs to remote lake restoration	1952 (lung) 1968 (lakes)	1979–2001 (increasing EU etc. restriction leading to c. 90% reduction of 1975 by 2010	27–58
MTBE in petrol as a substitute for lead	1960 (taste/odor/in water)	2000 (undesirable in Denmark/California, permitted elsewhere)	40+
Great Lakes contamination	1962/3	1970s (DDT banned in US and EU, 2000 debates continue regarding health-damagin pollution)	10–?
TBT antifoulants: a tale of ships, snails and imposex	1976–81 French oysters collapse	1982–87 (French, UK, then NE Atlantic ban; 2008 global ban	6–32
Beef hormones as growth promoters	1972/3 (estrogen effects on wildlife)	1988 (EU ban, US continues)	16+
Mad cow disease: reassurances undermined precaution	1979–86	1989 partial ban; 1996 total ban)	10–17

Above chart is based on Table 1 by David Gee in Pathophysiology 16:217-231, 2009.

Chapter 10

EMF SAFETY STANDARDS

Scientist James Lovelock is widely known as the creator of Gaia theory, which poses that living organisms and their inorganic surroundings have evolved together as a single living system. Less well known about Lovelock is that he invented a device called an *electron capture detector* (ECD), which "is the most sensitive, easily portable and inexpensive analytical apparatus capable of detecting substances present in the atmosphere at concentrations as low as parts per trillion."[1] The ECD's design went through several phases, beginning around 1948 until it was completed in 1959. Today, over 50 years later, the ECD remains the most effective tool for detecting pollutants in the atmosphere.

One day, while on a research expedition in Ireland, Lovelock turned on one of his early ECDs and found surprising results—an unexpectedly high level of chlorofluorocarbons (CFCs) in the air.

CFCs are organic compounds composed of carbon, hydrogen, chlorine, and fluorine that have been in many products from aerosols to refrigerators (DuPont's well-known chemicals, Freons, are CFCs). Today, we understand the various characteristics of CFCs that make it such a destructive force to earth's ozone layer (accelerating the greenhouse effect and the rates of global climate change). In the mid-1970s, however, the research on CFCs was limited. CFCs were considered to be a miracle substance, given the variety of their useful applications (including refrigeration, aerosols, and firefighting, among many others) and their relatively low cost of production.

In subsequent expeditions, Lovelock detected CFCs in Antarctica and the Arctic—further confirming its prevalence. After learning of Lovelock's work, Drs. F. Sherwood Rowland and Mario Molina, two researchers at the University of California at Irvine, investigated the potential impact of CFCs on earth's atmosphere. They reported in 1974 that strong UV rays could break down CFCs, releasing large amounts of chlorine into the upper stratosphere. Given that chlorine destroys ozone, these and other researchers hypothesized that increasing amounts of CFCs in the atmosphere would break down the earth's protective ozone layer—setting the stage for the atmospheric greenhouse effect and leading to global climate change. The team of Rowland and Molina was awarded the 1995 Nobel Prize in Chemistry (along with Paul J. Crutzen for separate research on the atmosphere).

Despite the alarming nature of these early findings, it would be another 12 years—and the 1985 discovery of a massive hole in the earth's ozone layer above Antarctica—before the world would react with the 1987 Montreal Protocol calling for dramatic reductions in worldwide production of CFCs, which, as of 2009, has been signed by all member countries of the United Nations.

Global action on this issue was delayed, in part, due to heavy industry lobbying. In 1975, DuPont, producer of 25% of the planet's CFCs, invested millions of dollars in a nationwide newspaper advertising campaign explaining that there was no proof linking CFCs to the destruction of the ozone layer.[2] A press release from the aerosol industry explained that the link between CFCs and ozone depletion was an unproven theory; this PR document was reprinted in the *New York Times, Wall Street Journal, Fortune* magazine, *Business Week*, and the London *Observer*.[3] In 1975, the CFC industry hired the world's largest PR firm, Hill Knowlton, to produce a speaking tour for Richard Scorer, a prominent British scientist and former editor of the *International Journal of Air Pollution*. Scorer used these talks to attack Molina and Rowland, explaining that "the only thing that has been accumulated so far is a number of theories."[4]

Leading CFC manufacturers warned of significant economic disruption resulting from a ban on CFC production. DuPont predicted that the costs of such a move could exceed $135 billion in the United States alone and that "entire industries could fold."[5] The CEO of Pennwalt, at the time the third largest producer of CFCs in the world, warned of "economic chaos" resulting from the cessation of CFC production.[6]

While such corporate opposition proved effective in delaying a global response to the CFC threat for over a decade, one country did take action shortly after Rowland and Molina's discovery. Despite the lack of overwhelming and conclusive scientific evidence, and in the face of increasing public concern, the US government determined the potential threat of CFCs to be so significant that immediate, unilateral action was required. As Donald Kennedy, then head of the US Food and Drug Administration (FDA), warned in 1978, ozone depletion "could increase the incidence of skin cancer worldwide, cause changes in the climate and have other undesirable effects."[7] That same year the FDA, the Environmental Protection Agency (EPA), and the Consumer Product Safety Commission made the United States the first nation in the world to regulate CFC production, completely banning their use in aerosol cans.

In this instance, regulators did not wait until the scientific information was more certain. They did not wait until irreversible damage had been done. They did not ask the industry to devise voluntary measures to reduce CFCs. They either did not believe the dire economic impact of the estimates generated by firms such as DuPont, or they believed the ozone layer to be worth more than the estimated costs of compliance.

And so, despite industry opposition, these US regulators practiced prudent precaution, banning the use of ozone-depleting CFCs in aerosols. Subsequent findings validated these preventative measures, and the rest of the world eventually caught up to the United States' environmental position, virtually eliminating CFC production worldwide. And, as we know, firms like DuPont (who produced CFCs) and Gillette (who used CFCs in their deodorant products) remain quite resilient businesses to this day.

Of course, CFCs and EMF are fundamentally different issues. While banning some substances, such as CFCs, is an option, banning EMFs is not. Replacements exist for CFCs in almost all uses, and thus regulating them out of existence does not present the same type of economic and social disruption as a ban on EMF emissions. After all, we still have not found a way to power a lightbulb, much less make a cell phone call, without generating electromagnetic radiation. Still, the regulatory acts taken by the United States in 1978 to reduce ozone depletion—before there was absolute scientific proof of danger to the planet or global consensus on the matter, and despite significant industry opposition—provide an instructive example as we approach the question of safety standards for non-ionizing electromagnetic radiation.

THE CHALLENGE OF ESTABLISHING EMF SAFETY STANDARDS

People have known for a long time that natural electricity can be dangerous. The ancient Greeks knew that the discharge from an electric eel could stun and sometimes kill swimmers. Humans have probably known for a much longer time that being hit by a lightning bolt could do the same, eventually leading to Benjamin Franklin's invention of the lightning rod. Once we learned how to generate electricity with machines and its use became widespread in our society, we have regulated its generation and distribution in order to protect ourselves and our property.

But assessing EMF safety and setting EMF standards is not a simple task. Even those who accept the science demonstrating bioeffects from EMF exposure explain that much more remains to be researched—that current science has raised more questions than answers. No one can say what constitutes an unhealthy "dose" of EMF because there are so many different physical aspects of the radiation to consider such as the voltage, the frequency, the pulse variations, and the duration of individual and cumulative exposures over a lifetime of using different devices, in addition to ambient natural radiation. There are also significant differences between individuals in the biological systems affected as well as in the ability to repair damage done by EMF. The information to date is insufficient to provide this level of detail about the potential health risks from EMF exposure.

LIMITATIONS OF SCIENCE

As discussed earlier, the two main types of scientific research informing the discussion on EMF bioeffects are epidemiology and laboratory science. In epidemiological research, we are limited by the inability to determine "proof" (recall, epidemiology can demonstrate correlation, not causation). Epidemiology allows us to say, as Interphone does, that cumulative use of cell phones greater than 1,640 minutes *correlates with* a 40% increased risk of developing certain types of brain tumors. But these results do not allow us to claim that 1,640 minutes of cell phone use directly *causes* a 40% increased risk of developing these tumors.

Another significant limitation of epidemiological studies is the difficulty of establishing true control groups. In research, we always need a control group, or a group of test subjects who can remain unexposed to whatever is being tested. It is close to impossible in today's world to establish a control group that is unexposed to man-made EMF in their daily lives. And even such rare individuals who are unexposed are not true controls, as they are exposed to many other forces and environmental stresses in their daily lives—too many to account for in research.

This is why we try to execute many epidemiological studies in addition to examining the biological data from high-quality laboratory research. However, epidemiological studies are not inexpensive to perform—especially when the health effects we are most interested in stem from long-term EMF exposure over periods of 25 years or longer. Studies shorter in duration are less costly to perform but shed no light on these long-term effects. (Research on the order of the 30-year COSMOS cohort study is exceedingly rare. It was launched in 2010 across six European countries—UK, Denmark, Sweden, Finland, the Netherlands, and France—and we shall have to wait over a generation for the results. During this time, many more people will be exposed to EMF from many more sources in addition to those in existence today.)

We might well wish to turn instead to the laboratory, where we can establish tight (though still imperfect) controls and quantify effects such as cellular damage from tightly regulated doses of EMF exposure. In general, laboratory studies of cell biochemistry and cell physiology have been quite successful in determining the biological processes that are activated on exposure to EMF from cell phones. However, while such experiments can inform our understanding of the biological mechanisms involved in any health effects, and while such studies can be useful in determining safety standards for human exposure, laboratory research cannot be effective in determining disease outcomes. You will note that none of the laboratory studies (primarily covered in chapter 4, as well as some in chapter 6 and 7) demonstrate that EMF causes cancer. They instead focus on effects from EMF exposure on very specific biological systems and pathways that can lead to cancer. But the relationship between these effects on systems in your body and possible long-term effects like cancer are unclear. As a result, one cannot draw conclusions regarding the question of human safety over extended periods from scientific research conducted in laboratories.

CONCLUSIONS FROM THE EMF SCIENCE

Neither epidemiology nor laboratory science provides us with a definitive answer to questions about EMF safety. Each approach has strengths and limitations. Results from both types of studies must be considered together. Even then, however, they do not provide a complete understanding of the subject. The National Institute of Environmental Health Sciences (NIEHS) at the National Institutes of Health (NIH) is not entirely incorrect to conclude that "the lack of connection between the human [epidemiological] data and the experimental [laboratory] data (animal and mechanistic) severely complicates the interpretation of these results."[8] However, they should also have pointed out that the laboratory data provide ample evidence of plausible and (in some cases) probable biological mechanisms that can account for the epidemiological data (e.g., DNA damage leading to mutations and the initiation of cancers).

In short, science has some critical limitations in its ability to help us answer questions such as will my cell phone give me cancer? or will my use of WiFi networks result in leukemia? Currently, science does not tell us, with any specificity, what health effects result from which types of EMF exposure. (And it is precisely this doubt on which the wireless industry depends in its continued efforts to forestall any possible regulation of their products.) What the EPA says in their 1992 "Questions and Answers about Electric and Magnetic Fields (EMFs)" remains true today: "The bottom line is there is no established cause and effect relationship between EMF exposure and cancer or other disease. For this reason, we cannot define a hazardous level of exposure."[9] The reader should be reminded that manufacturers were not required to show that their products were safe before being allowed to sell them to the public. These same data would not have allowed them to be sold to the public, precisely because EMF cannot be proven to be safe. Unfortunately, agencies set up to protect the public have not made sure about the safety of products before allowing their release.

It appears that the response of governments and industry groups to this lack of specific cause-and-effect relationships between nonthermal exposures to EMF and negative health effects has been to formulate regulations and safety standards that ignore them completely. While there is plenty of science indicating the presence of significant health risks at nonthermal levels, as far as safety standards and regulatory frameworks are concerned, EMF is harmful to humans only at levels powerful enough to result in increased temperature (the so-called thermal effect). No recognition at all is given to any potential health effects at lower, nonthermal levels of non-ionizing electromagnetic radiation, even though nonthermal biological effects have been scientifically demonstrated for over a century. It should be noted that among the nonthermal biological effects is the cellular stress response (i.e., the synthesis of stress proteins), a protective mechanism activated by exposure to a variety of harmful agents.

ICNIRP AND IEEE

The majority of safety standards established around the world to regulate non-ionizing EMF radiation are based on two sets of recommendations. One, from the International Commission on Non-Ionizing Radiation Protection (ICNIRP), was first released in 1993 and has been updated most recently in 2010. The ICNIRP exposure guidelines "were designed to evaluate the credibility of the various reported findings." The guidelines continue:

> Only established effects were used as the basis for the proposed exposure restrictions. Induction of cancer from long-term EMF exposure was not considered to be established, and so these *guidelines are based on short-term, immediate health effects* such as stimulation of peripheral nerves and muscles, shocks and burns caused by touching conducting objects, and elevated tissue temperatures resulting from absorption of energy during exposure to EMF [emphasis added]."[10]

In other words, per ICNIRP, exposure to nonthermal levels of non-ionizing radiation levels are safe, because they do not cause immediate, short-term damage (though, as we've discussed, they do lead to decreased production of melatonin and sperm cells). As the World Health Organization (WHO) summarizes in their endorsement of these recommendations, "EMF exposures below the limits recommended in the ICNIRP international guidelines do not appear to have any known consequence on health."[11] Have they overlooked effects on melatonin and sperm?

The other set of recommendations commonly used as the basis for EMF exposure safety standards was issued by the Institute of Electrical and Electronics Engineers (IEEE) in 2002. As with ICNIRP's recommendations, IEEE based their guidelines on EMF exposure levels with established, short-term health effects. The assumption of both of these organizations is that levels of non-ionizing EMF radiation that do not cause heating in human tissue are safe.

Since virtually every regulatory framework in the world pertaining to non-ionizing EMF exposure is based on one of these two sets of recommendations, consumers and citizens are not protected at all against the types of biological and health effects discussed in this book that are demonstrated to result from exposure to nonthermal levels of EM radiation.

Both ICNIRP and IEEE issue separate recommendations for power-line ELF, mobile phone radiation, and RF/MW radiation from cell phone antennas. These organizations also issue different guidelines for the general public and for high-EMF exposure occupations. These recommendations are included, for reference, below (see p. 164), primarily for the purposes of comparison to radiation levels stated in other chapters. Even without trying to understand or interpret the values in the tables, it is easy to see that these supposedly "safe" levels are much higher—*up to thousands of times higher*—than those that have demonstrated negative health effects in scientific studies cited in this book.

CURRENT APPROACH TO SAFETY IS FLAWED

While the science to date on this subject has left important questions unanswered, it has also unambiguously indicated that our current approach to EMF safety standards is fundamentally flawed. All product safety regulations for wireless communication devices pertaining to human health are based on a single incorrect assumption: that any potential damage results from immediate-term thermal effects of EMF radiation from single sources. Current regulations are intended to protect against, for example, your cell phone getting hot enough to damage your cells and to prevent you from getting a lethal shock from your electrical kitchen appliances.

It's a good thing that the thermal effects of EMF radiation are regulated. Thermal effects are real and dangerous. But these regulations protect you from only the short-term effects of excessive, thermal levels of EM exposure sufficient to create a heating effect in human tissue. As mentioned throughout this book, the science clearly demonstrates that there are biological effects from EMF exposure even at nonthermal levels.

It has been shown that EMF stimuli in the power-frequency (ELF) range and the radio frequency/microwave (RF/MW) range evoke the same cellular stress response even though they differ in energy by many orders of magnitude. This clearly shows that the energy level of the radiation is not a critical factor in the stimulation of stress-protein synthesis, a fundamental cellular protective reaction. The fundamental biological reactions to EMF appear consistent across the EM spectrum, even though the EMF energy varies. As the *BioInitiative Report* concludes:

> The effects of long-term exposure to wireless technologies including emissions from cell phones and other personal devices, and from whole-body exposure to RF transmissions from cell towers and antennas is simply not known yet with certainty. However, the body of evidence at hand suggests that bioeffects and health impacts can and do occur at exquisitely low exposure levels: *levels that can be thousands of times below public safety limits* [emphasis added].[12]

US PUBLIC SAFETY LIMITS

While the majority of EMF regulations around the world are based on the ICNIRP or IEEE recommendations, the actual implementation is confusing and haphazard. Tables of these values are given below.

ELF Exposure Limits

		Magnetic Field	**Electric Field**
ICNIRP	General	2,000 mG	5,000 V/m
	Occupational	10,000 mG	10,000 V/m
IEEE	General	9,040 mG	5,000 V/m
	Occupational	27,100 mG	20,000 V/m

RF Radiation Exposure Limits in SAR Units

Frequency	ICNIRP	IEEE
935 MHz	0.08 W/kg	0.4 W/kg
1,800 MHz	0.08 W/kg	0.4 W/kg
2,400 MHz	0.08 W/kg	0.4 W/kg

RF Radiation Exposure Limits in Power Density Units

Frequency	ICNIRP	IEEE
935 MHz	470 μW/cm²	623 μW/cm²
1,800 MHz	900 μW/cm²	1,200 μW/cm²
2,400 MHz	1,800 μW/cm²	1,600 μW/cm²

We are exposed to EMF from many different sources at home, at work, and in public. Different agencies and organizations regulate different devices and exposures from those devices, while some EMF exposures are entirely unregulated. For example, in the United States, while there are regulations on EMF emissions from cell phones, per the EPA, "there are no federal standards limiting occupational or residential exposure to power line EMF."[13] And thus, a huge source of EMF radiation, emitted in even very remote areas of the nation that do not have cell service or radio-station reception is completely unregulated.

Cell phone emissions are regulated based on how much energy a human absorbs while using the device—the metric discussed earlier in this book called the *specific absorption rate* (SAR). Based largely on recommendations from the IEEE, the Federal Communications Commission (FCC, which regulates devices that emit EMF radiation in the RF/MW ranges) has established permissible SAR levels based entirely around thermal effects, identifying 1.6 watts per kg (W/kg) of tissue as the maximum level.[14]

According to the FCC:

> At relatively low levels of exposure to RF radiation, i.e., levels lower than those that would produce significant heating, the evidence for production of harmful biological effects is ambiguous and unproven . . . A number of reports have appeared in the scientific literature describing the observation of a range of biological effects resulting from exposure to low levels of RF energy. However, in most cases, further experimental research has been unable to reproduce these effects. Furthermore . . . there has been no determination that such effects constitute a human health hazard.[15]

It should be noted that the cellular stress response has been reported across a wide range of ELF and RF frequencies, and that it is the well-documented reaction of cells to potentially harmful environmental stimuli (e.g., temperature).

With regard to broadcast antennas, the FCC's permissible levels are based on a similar standard, and they conclude:

> Public access to broadcasting antennas is normally restricted so that individuals cannot be exposed to high-level fields that might exist near antennas. Measurements made by the FCC, EPA and others have shown that ambient RF radiation levels in inhabited areas near broadcasting facilities are typically well below the exposure levels recommended by current standards and guidelines.[16]

Similarly, the FCC concludes that the increasingly common smart power meters (which use RF to communicate power usage to utility companies) are safe:

> The FCC standard provides a currently accepted factor of safety against known thermally induced health impacts of smart meters and other electronic devices in the same range of RF emissions. According to the FCC, exposure levels from smart meters are well below the thresholds for such effects.[17]

In the United States, the FDA also has authority (stemming from the Radiation Control for Health and Safety Act of 1968) over products that emit EMF to ensure their suitability for human usage. The FDA's mandate is broader than the FCC's, since the FDA oversees products that emit radiation throughout the EM spectrum, including medical and scientific equipment such as X-rays and germicide lamps, as well as cordless and cellular telephones.[18] The standards, administered by the FDA's Center for Devices and Radiological Health, are again based on thermal benchmarks on the assumption that nonthermal effects do not exist.

The federal government does recognize that workers in some specific careers or workplaces have higher, and possibly less safe, levels of exposure than average individuals. The National Institute for Occupational Safety and Health (NIOSH), a division of the Center for Disease Control (CDC), as well as the Occupational Safety and Health Administration (OSHA) have issued recommendations for workplace EMF exposure. (See Table on following page.) However, according to the CDC, NIOSH "and other government agencies do not consider EMFs a proven health hazard" and "because of the scientific uncertainty, no Federal limits for worker exposures to EMFs have been recommended or established in the United States."[19] The situation is slightly different at the state level, where 25 states, Puerto Rico, and the US Virgin Islands have implemented OSHA-approved regulatory frameworks for occupational EMF exposure.[20]

EVIDENCE FOR CONCERN

While science has thus far been unable to *prove* that exposure to non-ionizing EM radiation causes cancer, it has provided plenty of evidence for concern. The latest laboratory research indicates that the basis for the safety standards recommended by ICNIRP and IEEE (namely, that biological effects do not result from exposure to nonthermal levels of non-ionizing EMF) are fundamentally flawed. This reliance on the thermal standard may have been an understandable position in the 1980s, when their work began, but far better information is available today to indicate harmful changes in cell physiology resulting from nonthermal radiation.

Average magnetic field exposures for various types of workers (in milligauss)

Type of Worker	Average Daily Median	Exposures Range
Workers on the job:		
Clerical workers without computers	0.5	0.2–2.0
Clerical workers with computers	1.2	0.5–4.5
Machinists	1.9	0.6–27.6
Electric line workers	2.5	0.5–34.8
Electricians	5.4	0.8–34.0
Welders	8.2	1.7–96.0
Workers off the job (home, travel, etc.)	0.9	0.3–3.7

NIOSH workplace exposure chart. From "EMFs in the Workplace," a NIOSH publication (no. 96–129), 1996, http://www.cdc.gov/niosh/docs/96-129/.

Although it is true that the energy of an electromagnetic wave increases with its frequency and higher-frequency ionizing EMF is more energetic than non-ionizing EMF (so that microwaves are more energetic than radio-frequency waves, which are more energetic than ELF waves), the level of energy absorbed by a human cell does not necessarily correlate with the biological response it will have. As discussed in chapter 4, reproducible laboratory science demonstrates that exposure to supposedly "safe" levels of non-ionizing EMF can cause DNA damage (types of damage that are demonstrated to lead to mutations and cancer) and lead cells to invoke the cellular stress response (a reaction to environmental stimuli your body perceives as dangerous). We know that the cellular stress response can help cope with short-term exposure to environmental stresses, but these systems are not designed to mitigate damage from high levels of exposure, or extended or repeated exposures. The cellular stress response and the synthesis of stress proteins is direct testimony of cellular damage from the cells themselves.

Beyond failing to consider bioeffects at nonthermal levels of EMF exposure, federal regulations consider only short-term exposure to EMF from a single source. The regulations fail to consider exposure to EM radiation from multiple simultaneous sources across the spectrum, such as occurs when you, for example, make a cell phone call (1) standing close to a wall with electrical power lines near a circuit breaker (2), with a smart meter on the outside (3), in a location covered by two or three WiFi networks (4), while someone nearby uses a microwave oven (5). Nor do such regulations consider the cumulative effects of long-term exposure over a period of years.

PRESUMPTION OF INNOCENCE DOES NOT WORK AS A PUBLIC HEALTH POLICY

The US regulatory framework for EMF exposure is incomplete (lacking, for example, any federal regulations on power-line emissions) and based on the insufficient goal of ensuring no single exposure results in a heat stress response. These regulations are based on faulty assumptions, failing to consider nonthermal health effects, multiple simultaneous exposures, and cumulative exposures over a lifetime.

The establishment of such regulations has been based on recommendations from international standards bodies that recognize only the validity of thermal effects from high levels of EMF exposure. These committees recognize that there is some science indicating the possibility of negative health effects at lower, nonthermal levels but insist that the prudent approach is to wait and see what the science bears out over time. In other words, low levels of non-ionizing electromagnetic radiation are "innocent until proven guilty."

While the presumption of innocence is invaluable to the American system of justice, the same approach does not make sense as a public health standard when the risk of irreversible damage is so high for so many. This entire framework regulating EMF emissions and exposures is fundamentally flawed and must be rebuilt, from the ground up.

Because of the wide range of biological systems affected, the wide range of frequencies that are biologically active, the low response thresholds, and the possibility of cumulative effects by repetitive stimulation, the exposure standards should be revised to take into account the guidance provided by the new findings, specifically:

> The importance of nonthermal mechanisms in assessing risk.
>
> Total cumulative exposure across the different divisions of the spectrum from multiple sources.
>
> The increasing EMF background radiation due to the proliferation of cell phone and broadcasting antennas, as well as many different electronic devices in the home and in the workplace.
>
> The most sensitive populations (usually children) must be afforded even greater protection.

Making these changes, however, requires a fundamental shift in the manner in which regulations are formulated for devices that emit non-ionizing electromagnetic radiation. As we will see in the next chapter, the Precautionary Principle, as devised by the environmental movement, presents a compelling alternative vision for how to view and manage the risk presented by our wireless devices.

Chapter 11

THE PRECAUTIONARY PRINCIPLE AND THE *BIOINITIATIVE REPORT*

By the early 1970s, the great forests of West Germany were dying. The Germans suspected the cause was industrial pollution that resulted from the tremendous post–World War II economic growth. Eventually, research studies would reveal the link between industrial pollution, acid rain, and deforestation. But in the face of potentially irreversible damage to an irreplaceable national treasure, the Germans decided to act *before they had definitive proof* by passing the groundbreaking Clean Air Act of 1974 to limit industrial emissions. In doing so, the Germans adopted a new approach to countering environmental risks. The subsequent decades have seen *Vorsorgeprinzip* (literally, the "precautionary principle") become an underlying principle of German environmental legislation.

THE PRECAUTIONARY PRINCIPLE

The Precautionary Principle instructs us that *in the face of serious threats, a lack of scientific certainty never justifies inaction*. As the United Nations–hosted 1992 Earth Summit explained in the Rio Declaration on Environment and Development, "where there are threats of serious or irreversible damage, lack of full scientific certainty shall not be used as a reason for postponing cost-effective measures to prevent environmental degradation."[1]

While, in a court of law, it is incumbent upon the prosecution to prove the guilt of the accused, the Precautionary Principle places responsibility on those who trade in, and profit from, the risky product to prove the safety of their product. As the 1998 Wingspread Consensus Statement on the Precautionary Principle (issued by the Science and

Environmental Health Network at the Wingspread Conference Center in Wisconsin) states, "when an activity raises threats of harm to human health or the environment, precautionary measures should be taken even if some cause and effect relationships are not fully established scientifically. In this context the proponent of an activity, rather than the public, should bear the burden of proof."[2]

The Rio Declaration and the Wingspread Statement are just two of the many different expressions of the Precautionary Principle. All share these key elements:

- Those in authority must anticipate harm before a harmful activity occurs.
- It is the responsibility of those performing an activity to show that the activity will not result in significant harm.
- Those in authority must act to introduce cost-effective control measures to prevent or minimize harm resulting from the activity, even in the absence of scientific certainty.
- The need for control measures increases with the degree of uncertainty and level of possible harm resulting from the activity.

The Precautionary Principle is a proactive environmental policy designed to protect citizens from potentially adverse environmental influences in the face of incomplete information about the risks these influences present. The estimated costs of immediate action must be compared with the estimated potential cost of inaction. If the potential cost of inaction is plausible, significant, and irreversible, the Precautionary Principle tells us to act.

In other words, the Precautionary Principle is how policy makers say it's better to be safe than sorry.

PRECAUTIONARY REGULATION

Recall from chapter 5 that during London's 1854 cholera outbreak the town council did not wait until scientific proof existed to conclusively link the Broad Street pump to the deaths. Instead, in an early application of the Precautionary Principle, the council acted immediately, removing the pump when reasonable evidence of a threat to public health was found (it was only later that the precise cause—an infected baby's diaper contaminating the pump—was identified).

Imagine if the Precautionary Principle had been applied to tobacco. How many lives would have been saved had the burden of proof been placed on the cigarette makers when smoking was first linked to lung cancer and other diseases? The same question could be asked of asbestos, PCBs, X-rays, and many other environmental pollutants.

As one approaches the question of regulating EMF emissions and exposure, the Precautionary Principle can provide an instructive perspective. The World Health Organization (WHO) supports the ICNIRP safety standards (discussed in the previous chapter) and discourages its member states from deviating from them, until and unless weight-of-evidence levels of exposure lower than the ICNIRP permits are demonstrated to result in adverse health effects. Specifically, WHO EMF standards state:

> The existence of biological effects and health hazards can only be established when research results are replicated in independent laboratories or supported by related studies. This is further strengthened when:
>
> - there is agreement with accepted scientific principles
> - the underlying mechanism is understood
> - a dose-response relationship can be determined.[3]

The Precautionary Principle indicates precisely the opposite!

The Precautionary Principle has been applied before to product regulation in the United States. The ban on CFCs in aerosols is just one example. The Endangered Species Act applies a standard of evidence that is less than scientific proof in order for the Fish and Wildlife Service to designate a species as endangered; after all, once we have definitive proof that a species is extinct, it's too late to prevent extinction.

The evidence assembled to date on the health risks of exposure to non-ionizing EMF radiation, along with the exponentially growing scope of EMF emissions in the environment, meets the standard for application of the Precautionary Principle in devising regulations for product emissions and human exposure. This belief spurred a group of scientists closely involved in research on biological effects of EMF and actively involved in the Bioelectromagnetic Society (BEMS) to conduct the grassroots project known as the *BioInitiative Report* (BIR). BEMS is the major international scientific society dedicated to this area of research.

THE PRECAUTIONARY PRINCIPLE FOR EMF

The BIR had its origins among members of the Bioelectromagnetics Society (BEMS), where I served as president in 2007. Many members participated in a 2006 symposium that I helped organize to introduce our membership to the Precautionary Principle. I contacted Professor Michael Kundi from the University of Vienna, who agreed to act as cochairman and also present a talk on epidemiology studies of various environmental pollutants as a context for EMF studies. For the final speaker, we chose Cindy Sage of Sage Associates, a well-known EMF consultant, who reported on practical applications of the Precautionary Principle.

The Symposium Summary
2006 Bioeletromagnetics Society
Minisymposium of EMF Research and the
Precautionary Principle

Chairman: Martin Blank
Cochairman: Michael Kundi

The Precautionary Principle is a proactive policy to protect citizens from potentially adverse environmental influences when information about the risks they present is incomplete. Generally, we rely upon epidemiology studies to provide information about risk, but the results are often incomplete and ambiguous. Given the high cost of both overprotection and underprotection, we should use all available information for estimating the potential risks to society. It is from this broader perspective that we consider what can be learned about the potential risks of exposure to electromagnetic fields from:

- Scientific mechanisms (physiological systems affected, biological thresholds, biological variability, etc.)
  Speaker: Martin Blank
- Responses to previous environmental and occupational hazards (smoking, asbestos, etc.)
  Speaker: Michael Kundi
- How the Precautionary Principle has been implemented in connection with EMF (e.g., Switzerland, Italy, etc.)
  Speaker: Cindy Sage

The minisymposium took place at the annual BEMS meeting in Cancun, Mexico. The schedule consisted of three 25-minute presentations (for descriptions, see the symposium summary above), each followed by 15 minutes of discussion. When the time allotted for discussion had expired and questions remained, we extended the session until the end of the day. Even then, the discussion wasn't finished, and following the symposium, the speakers were joined by a small group who talked on through dinner and into the night. We then scheduled an additional session for the following day—and once again, the time ended long before the discussion did.

From the scheduled talks and from the discussions, we had learned that:

- Safety standards built around protecting humans from thermal effects of EMF radiation completely fail to con-

sider the many fundamental biological processes, well documented to be affected by EMF at nonthermal levels.
- The energy thresholds for biological damage are very low, and so the thresholds for potentially negative health effects are probably also very low.
- Simultaneous exposure to different frequency ranges could have additive effects on the exposed humans; similarly, effects of cumulative exposures must also be considered.

Further, we realized that the above-cited fundamental flaws in our approach to EMF safety cannot be addressed by tweaks or adjustments to current regulations. Instead, the entire approach to regulation must be reconsidered. The membership's interest in this topic was clear, as was their desire to use their expertise on EMF issues to inform a wider audience of their assessment. Those of us who had participated in the symposium and the discussion realized that something had to be done, and that we were the ones who could start the process.

THE *BIOINITIATIVE REPORT*

The spark set off at the symposium led the participants to form the BioInitiative Working Group that eventually organized the *BioInitiative Report* (BIR). The BIR (which you can download and read at http://www.bioinitiative.org) reviewed a wide collection of scientific evidence obtained in connection with studies of biological effects of EMF. The data was primarily focused on studies of RF/MW exposure (which are rapidly increasing), but also included studies of power-line ELF. The studies included both laboratory results as well as epidemiological research. Over 2,000 references were reviewed and listed, including results that indicated biological and health effects, as well as results that did not. It should be emphasized that (unlike many of the committees that were critical of the BIR) the authors of the BIR reports were scientists who were involved in the research they were reviewing, and they also included three presidents of the Bioelectromagnetics Society.

As the BIR explains:

> This Report is the product of an international research and public policy initiative to give an overview of what is known of biological effects that occur at low-intensity EMFs exposures (for both radio-frequency radiation RF and power-frequency ELF, and various forms of combined exposures that are now known to be bioactive). The Report examines the research and current standards and finds that these standards are far from adequate to protect public health.
>
> Recognizing that other bodies in the United States, United Kingdom, Australia, many European Union and eastern European countries as well as the World Health Organization are actively debating this topic, the BioInitiative Working Group has conducted an independent science and public health policy review process. The report presents solid science on this issue, and makes recommendations to decision-makers and the public.

The report was edited by Cindy Sage and David Carpenter, and was published online in August 2007. It was updated most recently in 2012. All contributors to the *BioInitiative Report* played their parts in helping to achieve the goal of reviewing and evaluating a wide range of the literature on EMF. However, I must single out Cindy Sage, who as coeditor, realized the importance of the goal from the beginning and shouldered the major burden of coordinating the effort (as well as a disproportionate and undeserved portion of the criticism targeted at the BIR, discussed below). Above all, she encouraged us to stick to the science as well as to our timetable, and she, together with coeditor David Carpenter, brought the project in on schedule.

BIR CONCLUSIONS

The BIR concludes, as has been described in this book, that there is substantial known and accepted science indicating biological effects resulting from low, nonthermal levels of non-ionizing electromagnetic radiation (levels currently considered safe by regulatory agencies). Among the documented damage resulting from EMF exposure in laboratory studies are genotoxic effects including DNA damage and DNA activation of the stress response, as well as adverse effects on immune function, neurology, human behavior, and melatonin production. The epidemiological studies included focus on brain tumors, acoustic neuromas, salivary gland tumors, leukemia, Alzheimer's disease, Lou Gehrig's disease, and breast cancer.

The BIR finds that "the existing ICNIRP and FCC limits for public and occupational exposure to ELF and RF are insufficiently protective of public health" and recommends that international agencies and organizations adopt the Precautionary Principle in establishing a new regulatory framework for EMF-generating technologies. In September 2008, the Parliament of the European Union agreed, citing the BIR when it decided by an overwhelming vote of 522 to 16 that the current EMF safety standards were obsolete and must be reviewed.[4]

The scientific papers included in the BIR were updated, and with several additional contributions, were peer reviewed and published in 2009 in a special issue devoted to EMF of *Pathophysiology*, a widely read and respected scientific journal. It covered the same general areas as the BIR, emphasizing molecular interactions with DNA and harmful effects on the function of the brain. Beyond the BIR, that journal issue also included articles on EMF effects on animals in the environment, effects on reproduction, and the surprising ability of human limbs to react to EMF signals. Additional evidence was presented from epidemiology and laboratory studies of significant biological effects due to EMF at levels far below the safety standards.

CRITICAL RESPONSES TO BIR

Despite a positive response to the BIR in many circles, there were some strongly negative reactions as well.

The BIR immediately generated criticism due to the manner in which it was released. Generally, scientific papers go through a formal publication process, in which the science, methods, and analysis of the paper are analyzed by other scientists before publication. The authors of the BIR felt the information to be so revelatory and the subject matter so pressing that the public should have immediate access to the report. This is why the authors, myself included, chose to publish online initially. Because this publication method did not include formal peer review, we organized a panel of prominent experts to review the BIR before it was released online and published their names along with the report on the website. Following this unconventional peer review prior to release, the BIR was later published as a conventionally peer-reviewed publication in the 2009 issue of *Pathophysiology*. The BIR papers are now a part of the regular scientific literature.

Other criticism of the BIR is based on the fact that the report's conclusions differ from the official recommendations of entities such as ICNIRP and IEEE. For example, in response to the BIR's conclusion that "RF exposures can be considered genotoxic (will damage DNA) under certain conditions of exposure, including exposure levels that are lower than existing safety limits,"[5] the IEEE's Committee on Man and Radiation (COMAR) responded that "this conclusion is inconsistent with the conclusions from weight-of-evidence assessments by the UK Independent Expert Group on Mobile Phones (IEGMP 2000), called the Stewart Report, and the U.S. National Research Council Expert Panel (NRC 2008)."[6]

IEEE's "weight-of-evidence" approach to evaluating the EMF science is the same method applied by the wireless industry itself, which sponsors so much of the research demonstrating no biological or health effects from low levels of EMF exposure that "outweighs" the science linking EMF to bioeffects. You will note that such objections do not dispute the validity of the science covered in the BIR, or the manner in which that science is covered—the IEEE criticism is, in essence, that the BIR should be disregarded because its conclusions are in disagreement with the scientific consensus of other groups.

The now-defunct Australian Centre for Radiofrequency Bioeffects Research (ACRBR) reached similar conclusions from their review of the BIR, noting that the authors of the BIR "each have [*sic*] a strong belief that does not accord with that of current scientific consensus."[7] The ACRBR continues: "This does not mean that what is written in the Report is invalid, but it means that we need to evaluate the content of the report itself." Such statements are intended to cast doubt on the scientific validity of the BIR, without actually criticizing or challenging any of the BIR's scientific coverage and conclusions. The ACRBR also initially objected to the fact that the BIR had not undergone peer review in the publication process.

Privately funded industry groups offered similar responses. The Electric Power Research Institute (EPRI) noted that the BIR's conclusions were not "cost-effective" and implied that the BIR should be dismissed given that its conclusions are at odds with prevailing wisdom:

Unlike the expert panels that conducted previous EMF health risk evaluations for IARC, the US National Institute of Environmental Health Sciences (NIEHS), and the World Health Organization (WHO), the BioInitiative Working Group was not convened by any governmental bodies or recognized health risk assessment organizations. Moreover, its conclusions, opinions, and recommendations are not consistent with those reached by previous panels.[8] I have no doubt that the expertise of the contributors to the BIR regarding biological effects of EMF exceeds that of most members of the expert panels mentioned.

Similarly, the Mobile Manufacturers' Forum noted that the recommendations from the BIR differed from "conclusions drawn by the 100-plus reviews, reports and government statements that have been published in this area from countries around the world."[9]

The Health Council of the Netherlands (HCN) had many complaints with the BIR, but to my reading, their primary disagreement was with the fundamental assumption that the Precautionary Principle should be applied to the issue of EMF emissions. Though this group acknowledges that "some experimental studies found indications that certain biological effects may occur upon exposure," the HCN adds that, "it is not known whether such effects may lead to health effects." It then concludes: "The BioInitiative report argues that any effect of electromagnetic fields on biological systems should be avoided, thereby ignoring the distinction between effect and damage. The Committee does not agree with this approach."[10]

That the conclusions of the BIR differ from those of the EMF standard-setting bodies is true—indeed, this was the primary reason the BIR was written. The spokesmen of industry and many scientists reject the idea of nonthermal effects from EMF exposure out of hand (just as doctors once scoffed at the notion that disease could result from invisibly tiny creatures called "germs" living on dirty hands). The BIR compiled a significant amount of evidence to the contrary. Instead of confronting this evidence, the BIR's critics tended to dismiss its findings without actually addressing them.

IMBALANCE

A common criticism of the BIR is that it lacks balance. For instance, the Dutch HCN stated that "the BioInitiative report is not an objective and balanced reflection of the current state of scientific knowledge."[11] Similarly, the French Agency for Environmental and Occupational Health Safety noted that "some sections do not present scientific data in a balanced fashion . . . and [the BIR] is written in militant style."[12] And the ACRBR commented that the BIR did not present "an objective and balanced reflection of the current state of scientific knowledge."[13]

These criticisms do not acknowledge that the BIR included data from over 2,000 studies and noted contrary findings among them. Many of the authors were active contributors to the research areas they wrote about, with hands-on experience—unlike many of the critics. Additionally, the BIR was subsequently updated and published in a peer-reviewed scientific journal—a process that underscores the scientific validity of the BIR's content and conclusions.

As mentioned earlier, the full text of the BIR is available at www.bioinitiative.org, and I recommend that all interested take the time to read the science. Do the BIR authors have opinions? Yes, most definitely. But this does not mean that the science itself, which has undergone peer review, is unbalanced. Whether the authors of the BIR are balanced or not does not alter the underlying science presented in the BIR in any way—or make the potential risks to humans and the environment any less dangerous. Given the histories of other products such as tobacco, mercury, asbestos, and lead, we know that the human and financial cost of ignoring the rising tide of scientific data rises with each passing day.

SO FAR, SO GOOD

As editor of the special edition of *Pathophysiology* devoted to EMF (published one year after the *BioInitiative Report*), I introduced the issue with a classic joke about our indifference in the face of overwhelming evidence of risk: A man has just fallen from the 86th floor of the Empire State Building in New York. As he passes the 30th floor, he is heard saying to himself, "So far, so good . . ."

I ended that piece with: "Overall, the scientific evidence shows that the risk to health is significant, and that to deny it is like being in free-fall and thinking 'so far, so good.' We must recognize that there is a potential health problem, and that we must begin to deal with it responsibly as individuals and as a society."[14]

There certainly is room for discussion regarding the specific levels of the standards to be adopted. But I flatly disagree with those who are unwilling to face the need for significant changes in our approach to EMF regulation following repeated scientific demonstrations of biological and health effects from EMF exposures—at levels far below existing standards. I am confident that time will reveal these individuals to be as wrongheaded and destructive to the public health as the defenders of cigarette smoking proved to be.

Application of the Precautionary Principle is one means that can help society address complex challenges such as the science underlying the health effects of electromagnetic radiation. Given the continued questions and uncertainty on the precise links between exposure to low-frequency EMF and negative health outcomes, regulations should be formulated out of an abundance of caution. The indications from known science are that existing safety standards are grossly inadequate, dating from a time when our understanding of this issue was far more limited than it is today.

What might application of the Precautionary Principle look like? Some specific steps, indicated by the European Environment Agency, include:

- Consumers, especially young adults and children (who are at the highest risk for brain tumors) should stick to texting and hands-free sets to avoid exposing their brains to EMR.
- Manufacturers should design hands-free phones that are easier for consumers to use.
- Cell phones should carry warning labels.
- Corporate-funded research needs to be more broadly focused on biological effects, rather than being limited to the "heating" effects of microwaves.
- Governments should place a research levy on cell phones to fund independent research.
- Governments need to better protect cell phone researchers from retaliation by the industry opponents.[15]

This list is focused on cell phones but one must also consider other EMF sources, such as WiFi in schools and smart meters, as well as usage behaviors. Accordingly, one could add many other recommendations, including:

- All cell phones should have a hardware-level button to enable and disable airplane mode (so turning off connectivity, and EMF transmissions, is a simple and quick action).
- All WiFi routers should have power switches, so they may be easily and rapidly turned off when their use is not required.
- Section 704 of the 1996 Telecommunications Act (TCA), which prohibits any restrictions on placement of cell towers due to health concerns, must be repealed.
- The minimum permitted distance between residences and sources of extremely high levels of EMF (such as

high-voltage power lines and cell phone towers) should be significantly increased.
- Electrical utilities should be required to run residential power lines underground to minimize ambient EMF emissions.
- Employees in certain high-risk careers (such as power-line workers) should be required to wear outfits made with EMF-resistant fabric while on the job.
- WiMAX installations should be banned in cities, and WiFi should be banned in primary and secondary schools.
- The rollout of smart meters should be prohibited.

It is important to remember that application of the Precautionary Principle is designed as an interim step—a stopgap measure. The risks of continued inaction are simply too great.

While many of us continue to push for improved regulatory scrutiny of EMF emissions, you should not wait for such action. As we will see in the next chapter, there are steps that you can and should take as an individual to minimize your exposure to potentially harmful EMF radiation, without going back to the stone age.

Chapter 12

MINIMIZING EMF RISK

A few years ago, I asked my power company to survey the EMF levels in my house, a typical suburban one-family home. When the engineer arrived with his meters, we started at the transformer on the power pole 300 feet away and walked back toward the house. As expected, the EMF levels decreased as we proceeded farther and farther away from the transformer. We then entered the house and began measuring levels inside. Once again, as expected, EMF radiation continued to decrease as we increased the distance from the transformer.

Then, as we approached the middle of my home, we were surprised to find increasing levels of EMF. We investigated and discovered that the elevated readings stemmed from a power distribution line located on the street running along the other side of my house (which was on a corner and thus close to power lines on two intersecting streets).

One can make educated guesses about one's exposure, but as the sources of EMF can often be obscured or hidden, it is simply impossible to know for sure without measuring. In approaching the task of minimizing one's risk of negative health effects from EMF, it is important to understand that EMF doesn't always come from obvious sources. It's not just cell phones or WiFi networks (though those are important). There are EMF-generating technologies all around you—particularly in urban and suburban areas.

A colleague in England found this to be true when he investigated the case of a child diagnosed with leukemia. It turned out that the wall along which the child's crib was located was immediately behind the circuit panel with the breakers for the home's power, which is a source of very high levels of ELF emissions. Taking measurements is the only way to really know what your exposure is. Fortunately, once you do, you can take steps to minimize your exposure.

WHY BOTHER?

Mark Twain is often credited with the wisecrack "there's no getting out of life alive." A former chairman of my department, a physician by training, restated the same idea (albeit with less humor) when he explained that "you've got to die of something." Living organisms have predefined limits. Disease and disability are part of life. We go about our lives exposed to risks. We usually try to minimize and control these risks, but we cannot avoid them entirely. Encountering forces that result in negative health effects in biological organisms is a part of everyday life.

Humanity today faces a significant number of challenges to the stability of the planet and modern civilization: population growth, air pollution, limited clean drinking water, deteriorating food and soil quality, massive deep water oil spills, a floating mass of discarded plastic in the Pacific Ocean the size of a small continent, rising sea levels, climate change—to name just a few. In the face of such overwhelming challenges, it is tempting to either ignore the risks or surrender to one's fate. This is especially true with the risks of EMF, which are invisible, odorless, poorly understood by the public, and generated by the tools and technologies that we love, and on which we rely to perform our basic everyday activities and functions.

However, the EMF issue is different from many of the other challenges we face as a species. Responding to climate change requires a global effort, as does creating a food-supply chain and water sources capable of supporting more than seven billion human beings. It is simply impractical to think that one can have a significant impact on any of these issues as a single individual.

EMF exposure, on the other hand, is more within our control. While the scale of EMF radiation in our atmosphere is massive, it also rapidly dissipates with distance from the many different sources, such that a four- or six-foot spread between you and the EMF source is often enough to significantly reduce your exposure. EMF (to our knowledge) does not disseminate and linger in the atmosphere (as do, for example, carbon emissions), impacting air quality thousands of miles away (though Milham's theory of dirty electricity does explain that some EMF emissions are conducted and spread over the power grid, far beyond the reach of the original source of the polluting radiation).

Similarly, EMF does not remain in landfills for decades—when the source is shut off, the EMF emissions immediately stop and disappear. Due to these characteristics of electromagnetic radiation, it is possible to significantly reduce one's exposure (leading to reduced risk of the negative health effects discussed in this book), without giving up technology or waiting for regulatory agencies to reformulate safety standards.

MINIMIZING EMF EXPOSURE

Contrary to the propaganda from the wireless industry, one need not be a Luddite to successfully minimize EMF exposure. The use of seat belts in automobiles is just one example of how society can reasonably approach risk management in daily life. There are about 40,000 fatalities on the road each year in the United States, but I won't give up driving my car. I have a car with antilock brakes and air bags, and when I get in, I fasten my seat belt and keep to the speed limit, thus reducing (though certainly not eliminating) my risk of dying or being injured in a car accident.

There are similarly simple ways to mitigate one's risk associated with EMF exposure. The two key principles for reducing risk are:

1. Minimize your use of EMF-generating technology.
2. Maximize the distance between you and those EMF sources when they are in use.

Reduce your EMF Exposure in every possible way!

- **Learn to recognize and measure EMF sources**
- **Stay as far away from them for as long as you can**

This advice is essentially an elaboration of the policy of Prudent Avoidance advocated by the US Congress's Office of Technology

Assessment and the Environmental Protection Agency (EPA) in the late 1980s and through the 1990s regarding exposure to power lines, cell phones, and cordless phones.[1] Prudent Avoidance recognizes that usage of EMF-generating technologies is increasingly unavoidable, and in the face of unknown potential health effects from them, it is prudent to avoid unnecessary exposure.

I wish to emphasize that we still do not know what a "safe" level of EMF exposure is (though, in some cases, we do know what an unsafe level of exposure is). Recall from earlier chapters that research studies have demonstrated biological effects of EMF exposure at even very low levels of radiation. Neither do we know what a "safe" distance is from these sources, which is another way of thinking about the level of radiation. Until these questions are answered, the only real rule is to keep as far away as possible. And until safety standards recognize nonthermal effects of EMF, you should not trust them or rely upon their assurances.

Prudent avoidance is possible when one knows the source of the EMF exposure. It is easy to reduce usage of the microwave oven and to ensure that you aren't in the kitchen when it's in use. Similarly, we can control how much time we spend on our cell or cordless phones.

But, as the example with which I started this chapter demonstrates, prudent avoidance is not always so simple and clear-cut. It's hard to avoid what one does not know exists. However, in many cases it is possible to overcome our inability to detect EMF with our senses by learning to use measurement tools.

MEASURING SOURCES OF EMF

Some measurement tools are very expensive, but simple (though somewhat less accurate) devices are satisfactory for the purposes discussed in this chapter. We only need a tool to detect the fields, to determine where they are high and whether we can lower the exposures with increased distance or EM-shielding.

You may actually own a crude RF meter already. If you still possess an old-fashioned portable AM radio, simply tune it to either end of the dial (where there are no stations broadcasting), raise the volume to the maximum, and walk around. Listen for audio static; you will notice that the static levels change as you move. That static is an audio measure of the ambient RF being picked up by the radio. The louder the static, the higher the RF being picked up around you.

To get more scientific measurements and objective readings of EM levels in the non-ionizing range of the EM spectrum, you can purchase or rent a relatively inexpensive electromagnetic (EM) meter. Of the different types of EM meters available, there are two that are relevant to the type of non-ionizing electromagnetic radiation discussed in this book.

The first is called a *gaussmeter*, which measures ELF radiation in the power-frequency range (which is emitted by power lines, the electrical wiring inside of your home, as well as the power sources for appliances and tools such as your microwave oven and hair dryer). When measuring EMFs from power lines, transformers, electrical substations, and appliances around the home, the gaussmeter should be set to read at 60 Hz (or 50 Hz, if you are in a country where that is the frequency used by the power grid) and be able to measure fields as low as 0.1 mG (milligauss). Many modern smart phones include these magnetic sensors (it's what powers the compass functionality included in many of these devices). If you have an iPhone or Android phone, you can download free apps that turn your phone into a gaussmeter (though there are reports of varying levels of accuracy with these tools). But to reiterate, gaussmeters measure only ELF from power lines and other AC sources—not RF/MW radiation from cell phones, WiFi networks, and other wireless communication.

In order to measure the level of radiation in the RF/MW range of the EM spectrum, we use a *power-density meter*, which detects EM radiation levels in the RF/MW range. (Some meters include both gauss readings for ELF radiation and power density readings for RF/MW radiation.) For our purposes, your power-density meter should be able to detect fields as low as 0.01 $\mu W/cm^2$ (microwatt per centimeter squared).

The detector in any EM meter is a wire coil. Because magnetic fields have a direction like a compass needle, the ability of the wire coil to detect a magnetic field changes as you tilt the device. Certain more expensive models include a triple-axis meter, which has three coils to compensate for this effect. I recommend purchasing a triple-coil version if funds permit. As with all new gadgets, you have to learn how to use them, but this is relatively easy for our purposes, where precise measurements are less important than qualitative assessments (to determine whether exposures are high or low).

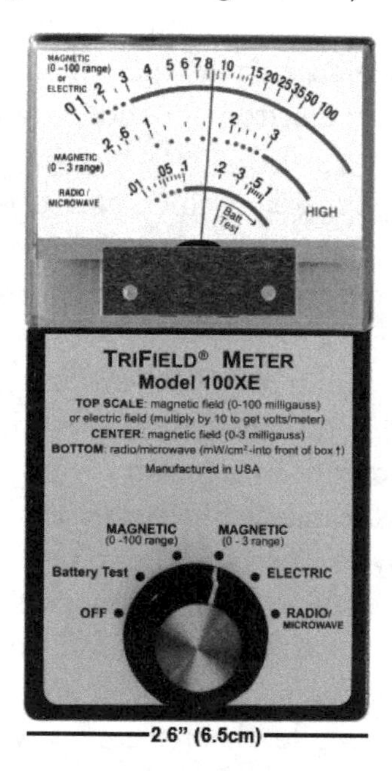

The versatile TriField meter, with its three scales (ELF electric field, ELF magnetic field, and RF electromagnetic field) is quite handy and sufficiently accurate.

It is always good to start by getting advice from people who know how to measure EMF. There are many sites on the Internet that provide useful information and also sell equipment. It is best to consult these sites and get a feel for what is available and how much the devices cost. Remember, you only need to get an idea of where the fields are and what their strengths are so that you can avoid them. There are many sites on the Internet to help you get started. I have found one URL that conveniently compares many types of EM meters: http://www.

emfcenter.com/metrsale.htm. Some other useful references include the following:

- The EMF Safety SuperStore (US): www.lessemf.com/
- Powerwatch (UK): www.powerwatch.org.uk/
- EMF Solutions (Canada): www.emfsolutions.ca/
- Canadian Initiative to Stop Wireless Electric and Electromagnetic Pollution: www.weepinitiative.org/

SOME BEST PRACTICES

To get really accurate measurements of your exposure, consider hiring a specialist. But there are a few guidelines to help you create better self-assessments using these consumer EM meters. First, as I said above, it is best to invest in a triple-coil gaussmeter for ELF; otherwise, your measurements will vary as you tilt the meter. (If you have a single-coil meter, the correct value will be the highest reading as you rotate the meter.) For RF, use a power-density meter.

You will want to ensure that you take measurements that reflect your daily habits. For example, if you are generally away from your home during the day, daytime measurements of your home may not reflect your actual exposure. It is possible, for example, that there is a nearby business that produces high levels of EMF from its machinery in the day and none at night. Instead, you will want to take the measurements at a time that matches your normal schedule.

You should take multiple measurements to accurately gauge your exposure. Cell phone companies, for example, can change their transmission frequencies during the day. Similarly, throughout the day, nearby WiFi networks can be turned on and off, and if you live in an apartment, EMF levels can shift frequently, as others in the building use different tools, devices, and appliances. Single measurements can help illuminate potential dangers, but extended exposure measurements are really required to gauge one's overall risk. The more measurements you take, the more accurate your assessment will be.

Finally, you will want to ensure that you measure EM levels from your own appliances when they are both on and off. Many devices these days are never truly turned off, but even if they are, these devices and their power cords can still conduct some electrical flow. Measuring EMF levels from devices when they are both on and off gives you a better idea of your exposures in daily life.

RISK-BENEFIT

Once you have a better understanding of your EMF exposure from the EM meter measurements, you can start to formulate a plan. Your plan should be based on a personal evaluation of the risks and benefits involved in the trade-offs: how much are you willing to change your behavior and at what perceived cost?

A variety of factors can influence such decisions. Nobel Prize–winning economist Daniel Kahneman expounds at length on this topic in his latest book, *Thinking, Fast and Slow*. Among the points Kahneman makes, media coverage heavily influences the public perception of risk. This is why, for example, tornadoes are seen as bigger killers (and, therefore, a bigger risk) than asthma, when the reverse is true; or why the public thinks that accidental death is more likely than death by diabetes, when again, the opposite is the case.[2] The more a risk is covered in the media, the greater the public's perception of that risk. As EMF, and the risks of EMF exposure, are chronic and therefore relatively sparsely covered in the media, this contributes to a public-wide underassessment of risk from electromagnetic radiation.

While media coverage is important, there are many other factors that should not, but do, influence human perception of risk. For example, Kahneman cites a study by Paul Slovic, a Professor of Psychology at the University of Oregon, in which the units of measurement used when relaying risks heavily influenced the public's perception of that risk (comparing, for example, "death per million people" and "death per million dollars of product produced"). As Slovic explains:

> "Risk" does not exist "out there," independent of our minds

> and culture, waiting to be measured. Human beings invented the concept of "risk" to help them understand and cope with the dangers and uncertainties of life. Although these dangers are real, there is no such thing as "real risk" or "objective risk."[3]

CONSUMPTION HABITS

Further complicating one's ability to perform rational risk-benefit calculations is our increasing reliance on the tools and technologies that are the source of bioactive EMF radiation. Humans didn't always find their way to restaurants with GPS maps on their iPhones, or shop online, or prepare their meals in microwave ovens. Just over a generation ago, in fact, none of these modern, EMF-fueled conveniences existed. And even just 20 years ago (when relatively few people had cell phones, the Web was only recently created, WiFi did not yet exist as a consumer product, and the tethered Internet was still primarily the domain of the military and academia), our daily behaviors were quite different, absent our persistently connected devices. Very quickly, technology has redefined the way we live, think, communicate, and interact.

Indeed, recent research from Dr. Betsy Sparrow at Columbia University reveals that our persistently connected world is literally altering how our brains work. In a paper entitled "Google Effects on Memory," Sparrow finds that we increasingly outsource our memory to the Internet:

> The results of four studies suggest that when faced with difficult questions, people are primed to think about computers and that when people expect to have future access to information, they have lower rates of recall of the information itself and enhanced recall instead for where to access it. The Internet has become a primary form of external or transactive memory, where information is stored collectively outside ourselves.[4]

Research such as Sparrow's indicates that people are becoming dependent on EMF-generating technologies for functions that had been previously handled by our own brains. And we increasingly see this in the behavior of many of our peers (especially, but not only, among younger age groups), as MIT professor Sherry Turkle describes:

> At home, families sit together, texting and reading e-mail. At work executives text during board meetings. We text (and shop and go on Facebook) during classes and when we're on dates. My students tell me about an important new skill: it involves maintaining eye contact with someone while you text someone else; it's hard, but it can be done.[5]

The behavioral dependence on cell phones has been termed *nomophobia* (a contraction of "no-mobile-phone phobia") and is characterized by compulsively checking your phone, worrying about losing it (even if it is in a safe place), and never turning it off. A 2012 survey in the UK revealed that 66% of respondents felt they had nomophobia (up from 53% just four years prior).[6] Respondents reported checking their phones an average of 34 times a day—and 75% report bringing their phones with them to the bathroom.[7] As we repeatedly check our mobile phones for messages, the habit transforms into an unconscious behavior. We hardly realize we are doing it, because it has become programmed into our reflexes.

Of course, the broad range of sociological, physiological, and neurological changes accompanying this evolution in our usage and consumption of tools are significant, and the subject of increasing amounts of research. However, from the perspective of EMF exposure, this is a concerning trend. This degree of reliance on the devices that generate EMF hampers one's ability to make rational decisions about EMF exposure.

PERSONAL CALCULATIONS

Each time you are exposed to man-made non-ionizing EMF that exposure carries a risk in the form of increased possibility of health problems in the medium to long term. Risk, of course, is only half of the risk-benefit calculation.

Some EMF exposures—such as from your neighbor's WiFi network, or a nearby cell tower—bring you no direct benefit at all. These may be easier to give up but are more difficult to avoid. Sometimes, though, we are exposed to EMF from a device or activity that does yield tangible benefits. For example, when you want to leave the office for the day and still be able to take your customers' calls, a cell phone is invaluable. In such instances, you should be able to place a value on the benefit received from the EMF exposure.

In my case, although there may be an increased risk of disease associated with the EMF radiating from my computer, I will not give up using it. It's a fantastic device that I use for work and for news, communication, and entertainment. I do, however, try to limit my exposure time, I refuse to use WiFi in my home, and I never keep my laptop on my lap for obvious reasons. On the other hand, I do not own a cell phone. My wife owns one that she keeps in the glove compartment of our car, but it is for emergency use only and is never on. We have a corded phone connected to a landline, and also use VoIP (Voice over Internet Protocol) services that have some of the advantages of mobile devices.

I could continue down a longer list of trade-offs and calculations that I have made, but you see where I am going. I have made my risk-benefit calculation, and I know what I am prepared to risk. But one size does not fit all. As with so many beloved habits that we know are bad for us—eating certain foods, for example—we must acknowledge the good with the bad. If too much of a good thing really can turn bad, laptops and cell phones must be used in moderation. Each individual must do a personal risk-benefit calculation, and communities must also start thinking the same way when they confront major changes associated with new wireless technologies, such as the introduction of WiFi in schools or the deployment of smart meters for measuring and reporting on energy use. Both of these relatively new technologies come with significant enhanced risk to the exposed populations, even though there may be other ways to achieve the benefits. For example, schools can use cables to connect to the Internet to avoid exposure. Regarding smart meters, the immediate benefits are to the companies who no longer have to hire meter readers.

EMF EXPOSURE MINIMIZATION TACTICS

Once you have considered the types of changes you are willing to make in your daily routines, here are some suggested tactics that you may adopt to minimize your personal EMF exposure.

Reduce Concurrent Exposures

An important consideration when calculating the risks and benefits of EMF exposure is concurrency of exposures—or simultaneous exposure from multiple sources. Virtually all of the scientific research into the bioeffects of EMF exposure focuses on a specific source of electromagnetic radiation. For example, some studies research the effects of ELF radiation, others research the effect of cell phones, and still others investigate the health impact of microwave ovens.

In real life, however, we are usually exposed to electromagnetic radiation across multiple frequencies in the EM spectrum. The science does not inform the question of health effects resulting from this type of concurrent exposure. I would recommend, per the Precautionary Principle, that it is much safer to assume that concurrent exposures bear unknown negative health outcomes that are likely to be additive. Concurrent exposures should definitely be considered and minimized. Thus, as just one example, if you do use a microwave oven, don't also use a cell phone at the same time; if you spend time on your cell phone, try to do so out of reach of a WiFi network; and so on.

Do Not Live Near High-Voltage Power Lines

There is no "safe" distance, though increased risk of health effects has been demonstrated at up to 1,200 feet from high-voltage power lines (generally transmitting 100–700 kV as opposed to 15–30 kV in distribution lines). If you live near sources of very high EMF, you should consider moving.

Keep Away from Transformers

The EMF near a transformer can be quite high, but the field strength diminishes quickly with distance and is generally not a source of concern. Some residences are, however, very close to these transformers. And remember, if you have a yard, take measurements there too (you may have a transformer closer to your yard than to your home).

Three transformers on a neighborhood power distribution line.

Live as Far from Cell Phone Antennas as Possible

Unlike the ELF transmissions from telltale cylindrical transformers on poles or high-voltage power lines, RF antennas are found in many different places and are frequently hidden or camouflaged. The EMF signals they emit are said to be very low, but that is a small comfort if the antenna is located on your apartment building's top floor, just above your bedroom (where you are continuously exposed as you sleep). Check your building and your neighborhood for these antennas and take a little time to measure the EMF coming from them.

Don't Use Electric Blankets and Heated Waterbeds

Electric blankets create a magnetic field that can penetrate about 15 cm

(6–7 inches) into the body and emit an electric field when connected, even when not actively heating. Epidemiological studies have linked exposures to electric blankets with miscarriages and childhood leukemia. Similar health effects have been determined for water bed heaters.

Changes in the wiring in recent models of these devices have minimized the EMF while maintaining the ability to heat. This was done by simply doubling the length of the wire carrying the current and having the wire double back on itself. This design results in two adjacent wires carrying current in opposite directions, so their magnetic fields are in opposite directions and tend to cancel each other out. This way of minimizing EMF by having the fields from adjacent cables interfere with each other is also used to design optimal arrangements for cables in power transmission lines. As a general precaution, disconnect even these improved electric blankets at night.

Run Extension Cords Clear and Away

Extension cords are very handy, but they generate ELF fields. You want to ensure that you run extension cords in locations away from furniture. Running an extension cord under your bed or couch will lead to much higher levels of ELF exposure to family members resting on the furniture. Ensure extension cords are laid straight and do not double back or loop, as such configurations can lead to unpredictably higher levels of ELF emissions. Do not cross other power cords with your extension cords; if you must, lay the cords at perpendicular angles.

Keep Cords Organized

Along the same lines, you should keep all of your power cords organized. When you use several separate electrical appliances and tools plugged into the same or nearby locations, it is easy for their cords to overlap or cross each other. As with extension cords, overlapping power cords can lead to unpredictably increased ELF emissions from the cords. Keeping them organized keeps these emissions more stable and predictable.

Switch from Electric to Battery Alarm Clocks

Electric clocks that plug into power outlets have very high magnetic fields—as much as 50 mG directly at the source and 30 mG one foot away, with high levels extending up to three feet away. These devices are particularly dangerous precisely because so many people sleep with them right next to the bed. If you use a bedside clock, you could be exposing your head to EMF equivalent to that of a neighborhood distribution line every night while you sleep. Switching to battery-powered alarm clocks eliminates ELF emissions, and in general it is wise to place all clocks (as well as all other electrical devices) at least six feet from your bed.

Efficiency Matters

Energy-efficient appliances are valuable because they consume less electricity to get the same amount of work done. Thus, they generate less pollution than older, less efficient models. This also means that energy-efficient appliances tend to generate less EMF radiation to accomplish the same amount of work as less efficient models.

Don't Use Fluorescent Lights

Fluorescent and compact-fluorescent lights produce much more EMF than incandescent bulbs. A typical fluorescent lamp on an office ceiling can have readings of 100 mG six inches away (by comparison, a single incandescent bulb can emit a field of 6 mG at a distance of six inches), but that decreases rapidly with distance. However, this is only a small part of the EMF dangers associated with fluorescent lamps. Unlike incandescent bulbs, in which the electric current causes a high-resistance wire filament to glow and emit light, fluorescents use high voltage to ionize the gas in the bulb and make it glow. Unfortunately, this adds RF frequencies to the EMF generated (generating dirty electricity, as discussed in chapter 3).

LED bulb technology generally emits lower levels of EMF radiation than fluorescent or incandescent bulbs, but in practice LED lamps powered by AC electricity can emit EMF with widely varying strengths, depending on how the lamp is wired.

Don't Use Dimmers, Three-Way Switches

Dimmers can be a nice feature in many homes, allowing inhabitants to set specific lighting levels depending on activity, time of day, and mood. Dimmers also generate a significant amount of additional ELF, due to their manipulation of voltage to provide variable levels of power to the bulbs. Remove all dimmers. If desired, replace them with three-stage lamps (which have different discrete levels of illumination, rather than the continuous spectrum of options afforded by a dimmer).

Three-way light switches (in which multiple switches control the same lighting fixture) can also generate significantly increased levels of ELF emissions, due to the fact that the wiring must often be installed in a configuration that enhances ELF fields within the home. Remove or disable such three-way switches.

Don't Use Radiant Electrical Floor Heating

Electrical heating by wires embedded in flooring can result in EMF levels over 100 mG at the floor and 30 mG at waist height! And, of course, that is throughout the entire area where the heating is installed. It is best to avoid these systems.

Don't Use Microwaves

There was a time not too long ago when everyone got by without microwave ovens. Sure, they make certain things very convenient. But it is worth noting that the safety limit for microwave leakage in the United States is at a power density of 5 mW/cm^2—about 500 to 5,000 times higher than in many European countries. But this is only part of the EMF from microwave ovens; the ovens also emit ELF fields of about 200 mG. Although the ELF exposure falls off rapidly with distance, the fields near microwave ovens are dangerous. Use microwaves very sparingly and from a healthy distance or not at all.

Take Care with Microwaves

Despite what I just wrote, many of you will no doubt continue to own and use microwave ovens because of their convenience. If you do, please do so responsibly. This largely means two things. First, get out of the kitchen when the microwave is on. EMF emissions drop off rapidly with distance, so the farther away one is from the microwave, the less EMF exposure results when the oven is on. And second, service your microwave. Despite their ubiquity, rarely do people have their microwave ovens serviced. The FDA's leakage limit of 5 mW/cm^2 is based on when the oven leaves the store. After months or years of usage, the oven will leak more and more microwave radiation. Proper servicing will reduce the emissions.

Position High EMF Appliances against Outer Walls

Many high-EMF-emitting devices are designed in a manner that the strongest electromagnetic fields emanate from the rear of the device. This is the case, for example, with refrigerators and many televisions. Place such major electrical appliances against outer walls, so as to not create EMF in any adjoining rooms in your residence. Of course, if you are in an apartment building, your outer wall may be someone else's bedroom or living room, and consideration should be afforded to your neighbors as well.

Airplane Mode Isn't Just for Airplanes

Most cell phones include an "airplane" mode where all wireless communication is disabled. This is for use in airplanes, so that the RF/MW radiation from passenger devices does not interfere with the airplane's equipment.

When not in airplane mode, your phone is in constant communication with network towers, continually transmitting RF/MW radiation (even more so if your smart phone has a WiFi connection, too). This is true regardless of whether you are on a telephone call. If you turn off your phone's data connections, however, the phone stops such emissions. When you don't need to make or receive calls, turn your phone into airplane mode.

And, of course, don't forget that you can also turn your cell phone off completely—which you should always do every night before bed.

Pockets Aren't for Cell Phones

Many people keep their cell phone in their pocket, particularly their pants pocket. I cannot emphasize enough how dangerous this is.

Remember: your cell phone is a microwave communication device, and unless it is turned off, it is constantly sending and receiving MW signals. Your pants pocket is very close to your reproductive organs. Place your phone in your laptop case, handbag, briefcase, or backpack—but never in your pocket, unless it is turned off.

SAR Is Virtually Useless

In chapter 3, I discuss many of the limitations of the SAR (specific absorption rate) measurement given by cell phone manufacturers and the FCC for all cell phones. As a result, this metric is essentially useless as an indication of safety and should never be relied upon as a basis for any decisions regarding personal health.

Not All Cordless Phones Are Made Equal

Cordless phones emit the same type of damaging radiation as cell phones. But cordless phones can be even worse than cell phones, as the base stations are located in the units that hold the receivers and therefore fill your homes with MW transmissions all the time. If at all possible, do not use cordless phones with your landlines. Alternatives exist, such as extended phone cords and extended wired headsets.

If you do use a cordless phone, realize that models with DECT (Digitally Enhanced Cordless Telecommunications) continuously radiate MW, whether or not the phone is in use. Try to buy a non-DECT model. And, I would also recommend buying a cordless phone that transmits over a lower frequency if possible (such as 900 MHz instead of 5.8 GHz).

Use a Headset—but Don't Delude Yourself

The cordless and cell phones discussed in chapter 3 are often used with headsets, some of which are wireless. Wireless headsets should be avoided, as they replace one microwave transmitter with another. Wired headsets help by increasing the distance between one's brain and the source of the microwave radiation. But this is only helpful if one keeps the phone away from one's body (and not, for example, in the front pocket of one's trousers). There are suspicions that wired headsets (which tend to dangle alongside the body when in use) may function as antennas for the EMF radiation generated by the phones to which they are connected, increasing the area of the body exposed to RF/MW radiation.

One can now find "airtube" headsets for sale on many websites. Such headsets rely on the vibrations of air, instead of a wire connected to a speaker. Many claim that such headsets reduce, or perhaps eliminate, the radiation to which one is exposed by standard headsets.

Unfortunately, the science is just not that well understood, and once again, the best advice is to minimize use of these wireless communication tools. Barring that, use a wired or airtube headset, but do not assume that this eliminates your MW exposure.

Laptops Tablets Aren't for Laps

Despite their name, laptops are not for laps. And despite Apple's advertisements with comfortable consumers lying on couches, browsing iPads nestled in their laps, neither are WiFi-enabled tablets. First, many laptops and tablets get very hot, leading to the type of thermal biological effects discussed earlier in this book and against which FCC safety standards are designed to protect. This heat may be acceptable if the laptop or tablet is on a table but not if the device is in your lap. If the computing device is running on battery power (which is DC), it is actually possible that it is not generating any noticeable levels of the type of EMF radiation discussed in this book. However, if your laptop is plugged into the wall (AC power) or if your tablet is plugged in through a docking station to wall power, the device generates ELF. Again, one does not want ELF radiation sources directly adjacent to one's reproductive organs. And, of course, all laptops and tablets have WiFi cards. If that card is enabled, the device emits microwave transmissions to communicate with the wireless router, regardless of whether plugged into the wall or running on battery power.

Computers and mobile computing devices are just not for laps. Like all EMF emitters, they should be kept as far away as possible when in use and they should be powered down when not in use. Similarly, WiFi cards should be disabled when not in use.

Evaluate Options for Electric Razors and Hair Dryers

As we saw in chapter 3, electric razors and hair dryers emit tremendous levels of EMF—up to 20,000 mG within four inches. Both are designed to be used in close proximity to the head, making it difficult to maximize the distance between the source and the brain. With electric razors, fortunately, there are options. If you use an electric razor like a clipper (to cut, rather than shave, one's hair), you can purchase a rechargeable razor. Rechargeable razors, unlike those plugged directly into the wall, generate DC fields, not AC. Alternatively, if you use an electric razor to shave, you can consider the new class of razors powered by AAA batteries.

Hair dryers are more difficult as there are no lower-powered replacements. Even if they did exist, their use would not be particularly optimal, especially for women who have long hair, live in cold climates, or have a schedule that does not permit enough time to allow their hair to dry naturally. While there are no models of hair dryers that emit anything close to "safe" levels of EMF, there are models of wall-mounted hair dryers that allow you to keep the motor (which is the source of the high ELF) at a safer distance. You want to look for a wall-mounted hair dryer in which the motor is part of the wall-mount; a vacuum-style hose then carries the heated air to your head. (In many wall-mounted hair dryers, such as those commonly seen in hotels, the motor is part of the extendable attachment, keeping the ELF emissions close to your head during use. This is to be avoided.) Another possibility is the Dryer Bonnet that fits over your head and is connected by a hose to the heater. In such devices, a *long hose keeps the EMF source away from your head.*

Unless you install a solution such as this, however, it is wise to minimize the use of hair dryers or not use them at all. And hair dryers should probably not be used at all on children as the high fields are held close to their rapidly developing brain and nervous system.

Ethernet Cables Still Exist

WiFi is convenient, but do you actually need it? Do you move your computer around a lot? Do you use a WiFi-enabled tablet at home? Or does your computer sit in one location? With the increasing power and speed and decreasing cost of WiFi technology (along with the absence of unsightly Ethernet cables), WiFi has become the de facto option for many home networks. But in many cases, such as individuals with desktop computers that tend to remain in a single location, relying on Ethernet instead of WiFi sacrifices no functionality.

Also, remember that Ethernet cables come in any length—up to hundreds of feet. Relying on Ethernet does not mean that you cannot get network access from multiple disparate locations in your house. Consider wiring your home with Ethernet (just as you would a telephone land-line) and using that in place of WiFi. You can still have a WiFi router that you plug in on occasion, when wireless Internet may actually be required.

Learn How to Use Your WiFi Router's "Off" Switch

For some reason, we've developed a habit of keeping WiFi networks on at all times of the day. In fact, I've seen many WiFi network routers that do not even have an "off" switch—these devices must be physically unplugged to be powered down. I know very few people who turn these networks off when they are not in use. However, when they are on, WiFi networks continually broadcast microwave radiation into our homes and offices. That makes for some very convenient Internet access—and it also makes for a lot of extraneous radiation exposure. Turn off your WiFi network when it is not in use.

Devices That Don't Turn Off Can Be Plugged Into Power Strips That Do

Many electronic products these days don't fully power down, even when you supposedly turn them off. However, such products easily plug into power strips that do have "on/off" switches. When a device is plugged into a power strip and the power strip is set to "off," your device gets no power. You can plug your WiFi router, your television, your cordless phone base, and any other devices without "off" switches into power strips and turn them all off at once—*completely* off—with the simple press of a button. Power strips are inexpensive and commonly available, and make turning off your devices (and ceasing their EMF emissions) quick and easy.

It's Smart to Avoid Smart Meters

Increasingly, power utilities around the country are encouraging the switch to smart-meter technology. Smart meters save on labor costs, eliminating the need for utility workers to visit your home to take readings of your power consumption. Smart meters are also much more efficient, enabling the utility to take measurements on a continuous basis. This, in turn, provides power companies with improved real-time data on power consumption across all of its customers, enabling them to make better decisions on how to manage the grid. And when a smart meter system, which includes an RF display unit connected to many appliances, is fully installed and enabled in a home, residents have access to data that allows them to monitor and regulate their own power consumption. The system also gives the power company the ability to turn off high-consuming devices during periods of peak consumption—again, helping to reduce energy consumption and increase the stability of the power grid. For these reasons, smart meters are widely perceived as a "green" technology.

However, this technology relies on RF transmissions. Precise levels of exposure from smart meters are difficult to ascertain since power companies do not typically provide clear information about how often the meters send RF transmissions (reportedly anywhere from every 30 seconds to every four hours). The power companies claim that the RF radiation from these devices is well within FCC limits. Those claims, however, are not independently verified. And even if they are correct, we've seen how FCC standards are insufficient to protect against disease resulting from long-term, repeated exposures. Smart meters are just one more source of RF radiation increasingly common in homes—and, unlike cell and cordless phones, smart meters are never turned off. Imagine your exposure if you live in a multiple dwelling and your apartment is close to where the meters are installed and transmitting regularly!

If your local utility or municipality is considering deploying smart meters, you should strongly consider participating in the opposition to the project. This is brand-new technology, and it is simply impossible to know the long-term health effects. These meters provide utilities with a significant savings in labor costs, but this comes at a significant risk to the users, who will be exposed to a radio transmitter operating around the clock on their home. Privacy issues have also been raised. In any case, the science we do know gives us significant cause for concern.

Switch to Plastic Eyeglass Frames and Foam Mattresses

Many metals conduct EMF and function as antennas. For this reason, if you wear eyeglasses, you should opt for plastic, as opposed to metal, frames. Metal eyeglass frames can serve as an antenna and focus radio and cellular phone waves directly into your brain. Similarly, it is likely safer to sleep on a foam mattress, instead of a traditional-style mattress, which includes metal springs that can act as antennas focusing EMF into the body during sleep. These springs may result in a double dose of broadcast EMF by adding a reflected beam of radiation.

EMF Exposure Minimization Tools

I have already discussed EM meters and certain other specific devices for purchase (such as the wall-mounted hair dryer, specially designed to reduce EMF exposures). By and large, however, the aforementioned tactics don't require the purchase of new gadgets. For those who are interested, however, there are products available for purchase to aid in the process of reducing one's EMF exposures.

I would be remiss if I did not mention that there are many sham products out there, marketed and sold by charlatans seeking to profit on fear, uncertainty, and doubt. Always examine the underlying scientific claims of such products. Can these companies point to specific claims (such as, "this product will reduce ELF fields by a specific percentage"), and are those claims backed by any research? If not, move along and do not buy. Fortunately, many sites (such as those URLs referenced earlier in the chapter) can provide guidance and assistance in your product search, helping to ensure that you purchase only quality products that do what they claim.

EMF Shielding

One popular set of products are those that shield EMF emissions. While magnetic fields can penetrate a lot of materials and substances, they cannot penetrate everything—it is possible to shield electromagnetic radiation. You can shield the source of the radiation (reducing the range of the EMF), or you can shield yourself (reducing absorption of EMF by the body).

If you Google "EMF shielding," you will see that many of the results are for scientific equipment. A lot of high-tech tools used by scientists and researchers are very sensitive to elevated levels of EMF, so EMF shields exist to protect this research equipment. These are generally different types of *Faraday cages*. A Faraday cage is a metal cage in which the holes in the wire mesh are of a specific size designed to repel specific frequencies of electromagnetic radiation. Faraday cages are likely impractical and/or undesirable for most of you; however, if you work in an environment with high levels of EMF generated by special equipment, you may ask your employer to install Faraday cages around that equipment for the health and benefit of the employees.

Other EM-shielding products are available to consumers. For example, you can purchase EM-shielding fabrics, which you can drape along walls (for example, if your neighbor's refrigerator backs onto your apartment) to suppress EMF emissions or to make clothing. You can also purchase premade clothing, such as hats and baseball caps, that repel RF EMF. Levi's line of Dockers clothing for men even sells a pair of pants with a "cell phone pocket" made of RF-repellent fabric. You can paint your walls, ceilings, and/or floors with EMF-shielding paint, which suppresses EMF emissions.

Please remember that all such shielding products only block particular frequencies of EMF—generally in either ELF or RF/MW frequencies. These shields cannot block all EMF.

Stetzer Filters

Graham-Stetzer Filters (commonly referred to as Stetzer Filters) are devices that filter out dirty electricity (discussed in chapter 3). Stetzer Filters do not eliminate ELF emissions from your electrical wiring, but they reduce your exposure to higher-frequency errant EMF emissions from the electrical wiring in your home or office. The efficacy of Stetzer Filters (regarding both the technical merits of the product, as well as demonstrated positive health outcomes from use of the filters) has been documented in peer-reviewed published studies by Dr. Sam Milham. For more information on Stetzer Filters, visit www.stetzerelectric.com.

Shielded Power Cables

As mentioned above, you want to use extension cords with the best insulation (i.e., shielded cables that can be grounded) to suppress ELF emissions. The same holds true for the power cords that plug into these extension cords, power strips, or directly into wall outlets. Measuring EMF levels from these cords will tell you which are the worst offenders. In such cases, you can attempt to shield the cords (using shielding products, such as those discussed above), or you can actually replace the cord entirely. Some EMF sites sell power cords with enhanced ELF shielding that can be used to replace those that come with your products. Because of the potential danger when working with such equipment, all such power cords should be installed by a licensed electrician.

At Work

At work, we experience many of the same kinds of exposure that we do at home from computers and appliances. Office machines, such as copiers and faxes, are also sources of repeated exposure. Virtually everyone is exposed to sources of dangerous EMF emissions in the workplace. But some individuals in specific careers face increased health risks. Electricians, power-line and cell-tower workers, welders, seamstresses, flight attendants, rail line workers—all of these are examples of careers with documented increased health risk stemming from exposure to increased levels of man-made and natural electromagnetic radiation.

Any risk-benefit calculation on personal EMF exposure must consider workplace exposure. But, of course, you have less control over your work environment than your home. While it may be tempting to simply accept the EMF levels at work, it never hurts to try to have them reduced. Oftentimes, this is simply a matter of education.

My Own Workplace Experience

When my laboratory at Columbia University Medical Center was moved to a different area in the same building, near the Facilities Management Department, I used my EM meters to measure the levels of electromagnetic radiation. I noted much higher levels than I had anticipated. I did a survey of the rooms in the immediate vicinity and found some unusual sources, such as a broadcasting antenna used to contact repair technicians anywhere in the building.

I spoke to the head of Facilities Management, who became interested in my measurements. I asked if I could do a quick survey of the area, and he agreed. There were indeed some hot spots in the area, especially where the ELF cable from the power company entered the building and also where the RF broadcasting antenna was located. I presented the results to him and explained the known science. Fortunately, he was interested and concerned. But at the time he didn't appear able, or willing, to do anything about the issues I raised.

About a year later, though, the University started a construction project during which personnel who worked in the area with high EMF readings were relocated and the area was converted into a storage unit. I met with the supervisor of that construction project, and he permitted me to do a follow-up EMF survey of the same space I had done earlier. The University had, in fact, lowered the EMF levels (probably by increasing insulation, rerouting cables, and other similar practices) and minimized the exposure of personnel in the area. I sent a report of my survey to the supervisor.

EMF Measurement Report, July 1, 2009

Here's a quick summary of the 60 Hz and RF (0.5 MHz–3 GHz) measurements that I made along the outer wall of the Black Building and in the basement hall. They are spot measurements and apt to vary during the course of the day, but I chose the same day of the week and roughly the same time. (Lunchtime should also be a relatively quiet time.)

I list three sets of measurements in order, going from the old Telecommunications Office end to the old Facilities Management end of the building, with one measurement about halfway in between. The values are listed as new/old (new being the measurements taken on Monday, June 29, 2009, at 12:30 pm and old being those taken on Monday, December 1, 2008, at 12:30 pm).

You'll be glad to know that all but one of the EMF values was reduced. (The single exception—2.8 / 1.9—is virtually identical.) As expected, the outer wall showed much greater mitigation of 60 Hz EMF, but the RF remains quite high, especially in the telecom corner.

Outer Wall	**TeleCom**	**Halfway**	**Facil.**
60 Hz (mG)	2.2 / 20	1.1 / 38	0.6 / 15
RF (μW/cm^2)	22 / 39	9 / 28	6 / 85

Hall	**TeleCom**	**Halfway**	**Facil.**
60 Hz (mG)	3.6 / 4.5	2.8 / 1.9	3.2 / 3.4
RF (μW/cm^2)	6 / 11	7 / 11	2 / 12

The University is almost always engaged in remodeling labs, so I'm not sure if there was any direct connection between my earlier report and the changes made during construction. However, the facilities did not appear to require renovation, and I would like to believe that once the authorities were informed, they did the right thing when they had an opportunity. Certainly, it can't hurt to approach your employers or coworkers with this type of information in a nonconfrontational manner in order to educate and inform them of the risks of EMF exposure. At best, you may end up making a significant reduction in your own exposures, as well as the electromagnetic radiation levels to which your coworkers are exposed. At worst, you will have helped inform others about the invisible and poorly understood risks of EMF exposure.

SUMMARY

I recommend that you practice prudent avoidance with all EMF-generating technologies. This certainly includes cell phones and WiFi, but also microwave ovens, hair dryers, televisions—every device that runs on AC power or that uses wireless communication as a source, and your exposure to that source should be minimized. Because of the rapid decrease in EM field strength corresponding to increased distance from the source of the radiation, stay as far away from the source of EMF radiation as possible in order to dramatically reduce your personal exposures.

The tactics covered in this chapter are not comprehensive, but instead represent a starting point for you to consider your own personal risk-benefit calculation when it comes to EMF exposure. As you start to think about the issue more and consider your personal exposures (as well as your relationships with EMF-generating devices), you will think of other options and alternatives to minimize your personal risk from the electromagnetic radiation that seems otherwise inseparable from modern civilization.

While prudent avoidance is a useful approach, there are groups of people who are far more vulnerable to EMF than most and for whom such practical techniques are insufficient. These individuals, discussed in the next chapter, must go to much greater lengths to reduce exposure to non-ionizing EM radiation.

Chapter 13

CHILDREN AND THE ELECTROHYPERSENSITIVE

While much of the research into the health effects of EMF exposure has focused on adults, EMF exposure first emerged as an environmental issue due to concerns regarding the health of children. In 1979, Nancy Wertheimer and Ed Leeper found a correlation between EMF exposure to ELF from power lines and the risk of childhood leukemia (discussed in chapter 5). Particularly disturbing was that the researchers noted the very low levels of EMF radiation that correlate with this form of childhood cancer. The study generated a lot of controversy and many follow-up studies, including research by scientist Dr. Sam Milham.

Milham's research (discussed in chapter 5) also observed a correlation between ELF exposure and childhood leukemia. Milham examined death certificates and other official records from across the United States at the time that the country's power grid was being built. He discovered that the pattern of childhood leukemia that we see today—with incidence peaking in children between the ages of 3 and 4 years old—corresponded with the introduction of electric power. Leukemia did not occur in children with these patterns until children were exposed to ELF radiation from the power grid; what's more, this pattern of childhood leukemia remains absent in those areas of the world, such as sub-Saharan Africa, where people are not exposed to ELF.

When all of the accumulated data on ELF and childhood leukemia were evaluated in 2002 by the International Agency for Research on Cancer (IARC) of the World Health Organization (WHO), the original conclusions of Wertheimer and Leeper were supported and power frequency magnetic field (ELF) exposure was ruled to be a "possible cause of cancer." In 2011, the IARC gave a similar evaluation regarding radio-frequency EMF and cancer, extending the warning over a wide range of non-ionizing radiation.[1]

CHILDREN AT GREATER RISK

Children's bodies operate differently than those of adults. In the context of our discussion, the most important difference is that children continuously grow at a rapid pace. The rate of growth in children means that they undergo a much faster pace of cell division. Thus, the DNA of children is more vulnerable to the errors that occur during normal protein synthesis, and any damaged DNA is more likely to pass to more cells (through cell division as well as replication), spreading further in the body and more rapidly.

Additionally, the bone in a child's head is thinner, leading to less shielding of the brain's neurons from external forces than is found in adults. Exacerbating matters further, research has shown that the amount of radiation absorbed by children from cell phones (the SAR values) is larger in children than adults, because children have higher levels of electrical conductivity than their older counterparts.[2] All told, the combined effects of these factors enable EM signals to penetrate deeper into the brains of children and affect DNA in more of their cells during cell replication and protein synthesis.

Other research from the University of Helsinki in Finland gives reason to suspect that children are susceptible to more subtle effects on brain function. These researchers demonstrated cognitive dysfunction in children resulting from exposure to cell phone radiation (902 MHz). Fifteen children were asked to perform auditory memory functions, with and without exposure to 902 MHz EM radiation. The results "suggest that EMF emitted by mobile phones has effects on brain oscillatory responses during cognitive processing in children."[3] In other words, exposure to cell phone radiation was shown to affect brain function in these children.

Any health risks created by exposure to non-ionizing electromagnetic radiation are going to be higher for children than for adults. (As discussed in chapter 9, the exclusion of the higher-risk population of children from the Interphone study was one of the points on which the research was heavily criticized.) Simply put, using the same cell phone, in the same position, for the same length of time does more harm to a child than to an adult. On this basis alone, it was entirely reasonable for the *Stewart Report* (published in 2000 by the Independent Expert Group on Mobile Phones at the request of England's minister for public health) to suggest limiting EMF exposure of British infants and children.

> Children may be more vulnerable because of their developing nervous system, the greater absorption of energy in the tissues of the head, and a longer lifetime of exposure. In line with our precautionary approach, at this time, we believe that the widespread use of mobile phones by children for non-essential calls should be discouraged. We also recommend that the mobile phone industry should refrain from promoting the use of mobile phones by children.[4]

It is simple and valuable advice: children should not use cell or cordless phones.

INCUBATORS

Children today are exposed not only to higher levels of EM radiation than any prior generation, but they are also exposed from an earlier age—with some children receiving very high levels of exposure right out of the womb. Preterm neonates and many sick infants are placed in incubators (that did not exist one or two generations ago) meant to provide the environment they need for survival and recovery. The equipment needed to maintain proper temperature control and air circulation produces EMF to which the children are continuously exposed. An additional EMF burden is generated by the instruments that monitor the physiological variables of the children and record them. All in all, the devices result in an EMF of about 12 mG at the top of the incubator mattress, many times above the field strengths of 3 to 4 mG that have been associated with an increased risk of leukemia.

A group of researchers in France, headed by Professor Carlo Valerio Bellieni, has been studying other effects of incubator EMF on physiological responses. They reported a temporary increase of approximately 15% in melatonin production shortly after moving newborns from incubators into standard cribs (where the EM fields were closer to normal ambient levels of about 0.1 mG).[5] Bellieni's results suggest that the low levels of EMF exposure from the incubator inhibited melatonin production in the infants placed inside—an effect that ceased almost immediately after the baby's removal from the device. Bellieni's findings also appear to corroborate the results of Joan Harland and Robert Liburdy, who demonstrated that comparable levels of EMF interfere with melatonin's ability to inhibit breast cancer cell growth.[6]

The life-saving impact of medical technology such as incubators cannot be denied. At the same time, one cannot ignore the potential health impact of resulting EM exposures to infants placed in these devices. Regulatory bodies must establish more stringent safety standards for EM exposure in incubators and other child-related technologies. There must be ways to achieve the same life-saving impact, while mitigating much of the risk stemming from the electromagnetic radiation inherent in the underlying technology.

BABY MONITORS

While many parents today cannot live without them, baby monitors are another relatively modern convenience. Baby monitors, like cordless phones, use digitally enhanced cordless telecommunication technology (or DECT) to continuously transmit signals like a cell phone. Most baby monitors pulsate microwave energy in the baby's room and beyond on a constant basis. Given that such monitors are generally positioned relatively close to the child, the baby's actual radiation exposure from a baby monitor is likely more than that from a nearby cell phone antenna (you would need to take measurements with an EM meter in order to evaluate the actual levels).

When Swiss Public Health tested two baby monitor units, they found the EMF to be well below the ICNIRP's 1998 SAR-value limit of 2 W/kg (it was actually only 0.5–4% of the 2 W/kg limit) and below 2 V/m (electric field) at 1 V/m. Regulatory agencies may deem these levels safe, but these exposures have been shown to be associated with significant physiological changes, such as the stimulation of protein synthesis during the cellular stress response. (It should be recalled from chapter 4 that the cellular stress response is the cell's reaction to potentially harmful stimuli.) The fact that these changes can occur indicates that the safety limits do not adequately protect children and that the devices cannot be claimed to be safe. I strongly advise against the use of such technology in the home—especially close to newborns and young babies.

PROTECTING CHILDREN

Children must be more strongly protected against EMF exposure than adults. Accordingly, all of the considerations and precautions mentioned in the previous chapter apply even more to children. While you should minimize the time you spend talking on your cell phone, children should avoid them altogether. While you may cut back on the number of days in the week in which you use a hair dryer, children should never use them at all. Never use the microwave when your child is in the kitchen. Never trust the reputed safety of a product that utilizes wireless radiation—as we've seen, such safety standards are inadequate for adults and are much less so with respect to the more sensitive and vulnerable population of children.

If children should not use cell phones, neither should they spend time in areas covered by WiFi networks. Of course, avoiding WiFi altogether is almost impossible given the plethora of hot spots all around us. But one can minimize exposure to WiFi networks by avoiding establishments that offer free WiFi access and turning off WiFi at home when not in use. However, children also spend a great deal of time in schools where, increasingly, they are radiated from school-wide WiFi networks that run all day. These can certainly be replaced by Ethernet connections that are well worth any additional cost.

From a technological perspective, the increasing availability of wireless connectivity in schools has great appeal. This technology is becoming cheaper every day, and there is no doubt that it can be used to enhance education. However, the price we will pay (through increased medical bills and earlier mortality) is much too high. This is especially so, because exposure can be avoided (recall, the Internet does still work over cables, too).

Just do the math: children go to school about 6 hours a day, 5 days a week, for about 36 weeks a year, for approximately 12 years (until the child reaches college, where they will also be exposed to ubiquitous WiFi). This equates to 12,960 hours of exposure to WiFi microwave radiation, for *each child*, by the time that child graduates high school—12,960 hours of exposure that could be avoided if the WiFi networks in schools are simply turned off and instead connected by cable.

And that's just the school day. One illustration of an increasingly common service is the deployment of WiFi in *school buses* in Vail, Arizona, to make the commute time more productive and less rowdy for students.[7] While the school reports that "behavioral problems have virtually disappeared," this policy also ensures that the children don't even get a break from exposure to microwave radiation when on the bus.

The rollout of WiFi in schools is still very new—too recent for any large-scale health impacts to have been reported. So the epidemiological science demonstrating health effects in children from exposure to WiFi does not yet exist and may not for decades. However, laboratory studies have clearly indicated potentially harmful changes that argue strongly for precautionary measures.

In the meantime, do you want your children to be the guinea pigs in this experiment to evaluate the health impact of near-constant exposure to microwave radiation throughout childhood? The rollout of citywide WiMAX networks, providing WiFi access throughout entire municipalities, must be reconsidered for the very same reasons.

The prospect of confronting a school system can often be intimidating to parents. But it is possible. And do not forget that the teachers are also affected by the same radiation!

STOPPING WIFI IN SCHOOL

I gave a talk in New York City on the increase in EMF in schools based on the scientific research I had done on EMF effects on DNA (summarized in chapter 4). One of the people I met there was particularly interested in what I had to say because she was concerned about the installation of WiFi in her neighborhood elementary school. She was worried that the process was being driven by marketing people and not by education experts. But what really set her off was the realization that no one had investigated the question of the safety of the children being exposed to the radiation from this new technology many hours a day during the school year. No studies had been done on children to determine safe levels; standards that had been set for adults were being applied to children without considering the enormous differences between adults and the young.

She had discussed this with other parents at her children's school and discovered that her doubts were shared by many. The parents as a group finally decided to bring their concerns to the attention of the school administration when the school announced that it was going to install a WiFi system. The reaction of these parents caused the school administration to realize that they had, at the very least, a public-relations problem. In response, the school organized a public-information meeting to let the engineering company responsible for installing the WiFi system present the project and answer questions. The parents requested that the program also include a presentation by a scientist on relevant research, but this was refused. The administration agreed that scientists could attend and take part in the discussion, but they would not be on the program. (This was another way of saying that the policy would be implemented because it was judged to be within the safety guidelines set by the authorities and that the meeting was only to inform the parents. In fact, by the time the meeting was arranged, the installation was already well under way.)

The parents group contacted me and asked me to attend this meeting to present the scientific evidence as the basis for the parents' concern about safety. I agreed that I could present the science, but that

I was not an engineer and could not comment on the technical aspects of the installation. (As it turned out, my rudimentary knowledge of the relevant engineering exceeded that of the representative of the company responsible for the installation.)

I arrived at the school early in order to make some measurements with my meters, one for measuring power-frequency EMF and the other for measuring radio-frequency power density. I performed a quick survey covering the outside perimeter of the school, the entrance, the auditorium, the hallways, and elsewhere—enough to determine the ambient levels and the pattern of possible exposure to both children and adults in the school.

The program started with the administration describing the installation as an important advance in technology that would greatly benefit the children's education. Children would learn new computer skills and be able to take advantage of educational programs on the Internet. The WiFi installers then gave a short description of the system, but they made the mistake of claiming that the fields were lower than the government safety standards—lower than what I had actually just measured.

At this point I entered the discussion and described the actual measurements I had just made and pointed to the instruments I had used. At the same time I mentioned that important biological changes had been documented at these low levels and that the parents had good reasons for their concerns about their children's safety. The contractors had apparently not made any measurements at all and had relied on manufacturers' estimates! They were clearly embarrassed.

This was enough to undermine the credibility and reliability of the contractors. The subsequent discussion was an exercise in damage control and attempts at face-saving by the installers and the administration. The school authorities tried their best to calm the parents and promised to examine the issue further. A few weeks later, the WiFi system was quietly removed, and the project was dropped.

These and other real-life experiences have given me some insight into how to approach similar situations to achieve the desired results:

- Avoid confrontation as much as possible—inform and try not to contradict.
- Trust but verify—take measurements when possible.
- Keep on teaching—you never know who is paying attention. I believe that as people become better informed about EMF, many will want to do the right thing and will want to apply pressure to ensure that governments and businesses do the same.

If you are a concerned parent (and especially if your child's school has yet to deploy WiFi—since it is generally much easier to prevent installation of a WiFi network than to have one removed), there are some organizations that can help guide you in this process:

1. www.magdahavas.com/category/electrosmog-exposure/schools/
2. www.safeschool.ca
3. www.citizensforsafetechnology.org
4. www.centerforsaferwireless.org
5. www.wiredchild.org

IN THE WOMB

Since infants and children are more susceptible to potential damage from EMF exposure, it makes sense that there could be similar, if not more significant, effects in embryos and fetuses. These possibilities were investigated by a team of researchers at the Kaiser Foundation Research Institute in Northern California led by Dr. De-Kun Li.[8] The research team outfitted 969 San Francisco–area women in their first 10 weeks of pregnancy with EMF meters for 24 hours (the usage of EMF meters ensured that the results would reflect accurate radiation measurements and not be subject to recall bias). Each of the subjects also kept a diary of their activities for that day.

Li then tracked the pregnancy outcomes, demonstrating that miscarriage risk increased when the mother was exposed to EMF of more than 16 milligauss (mG) for even a short period. These high-exposure women were nearly twice as likely to miscarry as the control group. From these data, it appeared that an acute large exposure could be the trigger for a miscarriage. The risk for miscarriage increased for all participants by an average of 80%, but it was higher for early miscarriages (220%) and among women with problems in prior pregnancies (400%).

Curious whether the mother's EM exposure during pregnancy had any correlation with rates of asthma in the children, Li continued tracking the children born in that study for a period of 13 years. There had been an unexplained sharp increase in asthma in recent years, and it turns out that there may be a relation to EMF. Children born to mothers who had an average EMF exposure greater than or equal to 2 mG during pregnancy were 3.5 times more likely to develop asthma by age 13. The risk of asthma in the child increased in a clear dose-response relationship with the mother's EM exposure, where "every 1-mG increase of maternal MF level during pregnancy was associated with a 15% increased rate of asthma in offspring."[9] Li's results (published in 2011) suggest that exposure in utero can cause damage to the developing fetus that manifests itself in later childhood as asthma. Because the continued rise in environmental EMF also means a rise in EMF exposure in utero, the results of this study could very well account for the recent increase in the incidence of asthma. The team has also reported a rise in childhood obesity correlated with maternal exposures exceeding 2.5 mG for 2.4 hours per day.

ELECTROHYPERSENSITIVITY (EHS)

The young are not the only population with increased vulnerability to the effects of non-ionizing electromagnetic radiation. Another population with unusual levels of sensitivity to EMF has recently come to our attention.

Most of the studies cited in this book indicate significant differences in impacts between different individuals (with overall averages being reported in the results). This type of data indicates that individual sensitivity to EMF can vary widely. Some individuals are significantly less sensitive and vulnerable than others, and a small segment of the population demonstrates physical responses when exposed to even extremely low levels of human-made EMF. This condition is known as *electrosensitivity* (ES) or, increasingly, *electrohypersensitivity* (EHS). (The term EHS implies that we are all ES to some extent.)

Though there is no specific source or amount of EMF that is known to trigger and foster EHS symptoms, this condition is recognized by the World Health Organization (WHO) as "a real and sometimes a disabling problem" for people whose "exposure is generally several orders of magnitude under the limits of internationally accepted standards." WHO defines the symptoms as including "headache, fatigue, stress, sleep disturbances, skin symptoms like prickling, burning sensations and rashes, pain and ache in muscles and many other health problems."[10] Professor Magda Havas from Trent University in Ontario, Canada, has shown that the heart rate in EHS people is greatly accelerated by relatively low levels of EMF. She believes the health problems can include

> cognitive dysfunction (memory, concentration, problem-solving); balance, dizziness vertigo; facial flushing, skin rash; chest pressure, rapid heart rate; depression, anxiety, irritability, frustration, temper; fatigue, poor sleep; body aches, headaches; ringing in the ear (tinnitus) and are consistent with chronic fatigue and fibromyalgia.[11]

The British group Powerwatch, which focuses on EMF and health issues, claims that between 3%–5% of Britons suffer from this condition[12] (a rate that would suggest more than 13 million sufferers across Europe), and that over 3% (approximately 250,000 Britons) report symptoms of EHS when in contact with certain EMF sources.[13]

Despite acceptance of EHS as a real and sometimes debilitating condition in some parts of the world, most countries have unfortunately tended to treat EHS as largely psychosomatic. This is despite an increasing number of documented examples of EHS. One prominently known sufferer of EHS is Dr. Gro Harlem Brundtland, the former prime minister of Norway and director-general of the World Health Organization (WHO). Brundtland was not born with EHS, but instead developed the condition following an accident with a microwave oven. As a result, her eye is now extremely sensitive to electromagnetic radiation, and she is certainly considered a reliable witness. The increasing body of scientific studies supports the experience of people like Brundtland and demonstrates that EHS is, indeed, a genuine condition.

In 1991, doctor and surgeon William J. Rea at the Environmental Health Center in Dallas performed one of these studies, exposing 100 patients, each of whom claimed to suffer from symptoms of EHS, to EM fields of 0 Hz (a "blank" exposure, for control) to 500 Hz. Prior to exposure, the researchers measured various bodily functions of the subjects, including blood pressure, pulse rate, respiratory rate, and temperature. If, following exposure, "the number and/or intensity of symptoms were 20% over baseline" in the subject, that subject was considered to have had a positive response for EHS.

One quarter of these patients was found to be sensitive to the EM-field exposure but not to the blank exposure. These 25 individuals with identified EHS were then twice again exposed to the same range of stimuli to which they responded in the first round. And in all three rounds of exposures, all 25 reacted to the same range of frequencies and all 25 failed to respond to the blank exposures. The authors concluded that they had demonstrated strong evidence of the existence of EHS as well as the ability to replicate it in laboratory conditions.[14]

Several studies of those who claim to suffer from EHS have identified objectively identifiable indicators in these sensitive individuals. Much of this research points, in particular, to differences in *mast cells.* Mast cells are present in many types of tissue in your body and are known to play a role in the activation of allergic responses. Between 1990 and 1995, Dr. Olle Johansson from Sweden's Karolinska Institute was involved in several studies that demonstrated that mast cells are present in elevated levels in the skin of EHS sufferers.[15] Johansson summarized these and other findings in 2010 when he published a review of published literature in this area, concluding that "it is evident from our preliminary data that various alterations are present in the electrohypersensitive persons' skin that are not indicated in the skin of normal healthy volunteers."[16] Studies such as Johansson's suggest explanations as well as a possible biological mechanism that could trigger the allergic responses in EHS sufferers.

While scientists like Johansson and Rea attempt to learn more about EHS inside the laboratory, outside of the laboratory the problem is growing. Citing a "dramatic rise" of symptoms in people using wireless technologies, a group of physicians in Germany have signed a document known as the "Freiburger Appeal." The most common symptoms identified in the document (that are often "miscategorized as psychosomatic") include the following:

- Headaches, migraines
- Chronic exhaustion
- Inner agitation
- Sleeplessness, daytime sleepiness
- Tinnitus
- Susceptibility to infection
- Nervous- and connective-tissue pains

EHS people do not all present the same symptoms, but they all do react to EMF in the environment, especially from wireless technologies. The strongly worded statement in the Freiburger Appeal calls for tighter regulation of wireless technologies[17] and, to date, has received the signatures of more than 3,000 doctors from around the globe. The symptoms of EHS can be so severe as to limit one's ability to work full-time and may even require medical care. In the most extreme cases, people try to relocate to EMF-free rural areas. One such place in the United States is Green Bank, West Virginia.

UNITED STATES NATIONAL RADIO QUIET ZONE

As discussed in the previous chapter, you can do a great deal to minimize—or in extreme cases, virtually eliminate—human-made EMF exposure in the home. However, as we've also noted, you are also exposed to human-made EMF emissions in the environment from sources such as cell phone and television broadcast antennas as well as WiFi and WiMAX networks. And while it may be easy for most of us to ignore this ambient radiation, it is a fact of modern life that EHS sufferers are too keenly aware of.

There are many rural and undeveloped areas of the country with relatively low levels of ambient EMF emissions. But then there is Green Bank, West Virginia, located in the heart of the 13,000 square-mile United States National Radio Quiet Zone, where regulations cap EMF emissions.

Established in 1958 by the Federal Communications Commission, the Radio Quiet Zone exists primarily to protect a very powerful radio telescope (there are also other, less significant installations of sensitive military and intelligence equipment protected by the zone). The Robert C. Byrd Green Bank Telescope is the world's premiere single-dish radio telescope operating at meter to millimeter wavelengths (0.1–116 GHz), a range larger than any other telescope. It is used to monitor EMF emissions from outer space for about 6,500 hours every year and for educational programs most of the remaining time. This telescope is extremely sensitive to RF radiation, and so to ensure its research can continue, the Radio Quiet Zone has restrictions on the EMF that can be generated and broadcast in the vicinity. As *Wired* reported in 2004:

> All major transmitters in the Zone are required to coordinate their operations with the national observatory. Radio stations point their antennas away and operate at reduced power. Cell phone base stations are few and far between, and entirely absent deep in the Zone. Even incidental electromagnetic emitters are regulated: Power lines must be buried 4 feet below ground. The wireless LAN card in your laptop? Forget about it.[18]

The Radio Quiet Zone may not have been designed as a refuge for EHS sufferers, but as the BBC reported in 2011,[19] dozens of EHS sufferers have moved to the region, which encompasses several cities in

Virginia and West Virginia. EHS sufferer Diane Schou moved to Green Bank, and her whole life changed:

> Living here allows me to be more of a normal person. I can be outdoors. I don't have to stay hidden in a [EMF-repelling] Faraday Cage. I can see the sunrise, I can see the stars at night, and I can be in the rain. [Living] here in Green Bank allows me to be with people. People here do not carry cell phones so I can socialize. I can go to church, I can attend some celebrations, I can be with people. I couldn't do that when I had to remain in the Faraday Cage.[20]

Not everyone can move to Green Bank. But as I said above, there are plenty of locations throughout the country with much lower levels of man-made electromagnetic radiation in the atmosphere. One of my sons lives in a very rural part of Oregon, 18 miles away from the nearest cell phone service—inputting his address into http://antennasearch.com shows zero antennas of any kind. It is not formally designated as a radio quiet zone. It's simply quite mountainous (impeding RF/MW signal distance), very rural, and undeveloped. Those characteristics mean that it is just not sufficiently profitable for cell phone companies to service this region of Oregon, and many (though a decreasing number of) similar regions around the country. Of course, most such regions still have ambient EMF radiation from radio stations and power lines. Still, those who live in such places have generally much lower exposures to human-made EMF in the environment than those who live in more populated areas. There are fewer wireless services (sometimes none, beyond radio), fewer power lines and transformers, and fewer neighbors and businesses with AC-powered appliances and tools.

Of course, such rural locations are not for everyone—the lifestyle is very different and there are fewer opportunities for work. Not all EHS sufferers (much less all pregnant women and children) can move to low-EMF zones. For those who cannot, it is even more important to implement lifestyle choices to minimize EMF exposure, following techniques such as those discussed in the previous chapter.

As I described earlier, the degree to which one is willing to reduce EMF exposure is based on a personal risk-benefit calculation. Diane Fox presents an extreme example of reducing EMF exposure. An EHS sufferer, Fox lives just outside the Radio Quiet Zone and claims that she is sensitive to even the extremely low ambient levels of man-made EMF in the region. In addition to living near the zone and obviously living without WiFi and cell phone, Fox has a propane-powered refrigerator, kerosene lamps, and a wood stove.[21] Again, while an extreme case, Fox provides an example of the possible ways in which individuals, particularly the most vulnerable, can reduce exposure to electromagnetic radiation in daily life.

CONCLUSION

Green Bank and other low-EMF zones aside, there are virtually no areas in our society that have remained untouched by the spread of EMF. Exposure to electromagnetic radiation has risk for all individuals, but as we've just discussed, fetuses, newborns, children, pregnant women, and those afflicted with EHS are particularly sensitive to the potential damage such non-ionizing radiation can cause. Those who are most vulnerable should take stronger precautions to minimize exposure. Instead of reducing time spent using a cell phone, such individuals should not use cell phones at all. The same holds true for microwave ovens, hair dryers, and WiFi networks.

After having discussed at length the harmful effects that can result from exposure to electromagnetic radiation, it may come as a surprise that the biological effects of non-ionizing EMF can be used to our advantage. As we will see in the next chapter, this is precisely what decades of widely accepted scientific research demonstrate.

Chapter 14

THERAPEUTIC USES OF EMF

Russia is not regarded as a particularly healthy place. Russians have a shorter life expectancy than citizens of any other industrialized nation. They have one of the world's highest rates of alcoholism and tobacco consumption. Their medical infrastructure lags significantly behind that of other advanced nations, in large part due to the legacy of neglect from decades of communist rule. But there is a device that is commonplace in Russian hospitals and ambulances that is frequently used to treat a wide variety of ailments with reportedly successful outcomes. According to a study of 3,000 Russian medical practitioners, this device, known as a *SCENAR*, is credited with:

- 79% improvement in the musculoskeletal system, muscle injuries, and diseases such as arthritis, sciatica, lumbago, and osteoporosis
- 82% improvement in many circulatory disorders, including strokes, thromboses, and heart failure
- 84% improvement in virtually any respiratory problems
- 93% improvement in both eye conditions and diseases of the digestive tract[1]

That's an impressive rate of success. And, notably, that success is reported across several of the body's different systems, including the cardiovascular, respiratory, skeletal, muscular, nervous, and digestive systems. Originally designed to help keep cosmonauts healthy while in space, SCENAR (Self-Controlled Energo Neuro Adaptive Regulation) detects the body's natural EMF emissions, notes where such levels are off, and then emits ELF radiation to stabilize the body's energy. In other words, SCENAR uses non-ionizing EMF radiation—the very same type of radiation discussed at length in this book as a public health hazard—to achieve high rates of success at healing the human body.

DOSE

When using a SCENAR device, one is exposed to a tightly controlled and regulated dose of EMF radiation designed to elicit a healing response. This is quite different from, for example, the level of radiation to which you are exposed when using a cell phone. Cell phone radiation is not designed to elicit healing responses in humans; indeed, it has been designed around safety standards that do not recognize *any* biological effects at all (beneficial or detrimental) of nonthermal levels of EMF radiation.

This notion of dose and response is critical in biology. There are some substances (such as potassium cyanide) that are toxic at more or less any dosage; similarly, there are other substances (such as water) that are healthy at very nearly any dosage. But there are many substances that are potentially dangerous at certain doses, while healthy, beneficial, or at least harmless at lower dosages. With EMF, the dose (more commonly referred to as EMF exposure) is important in assessing the resulting biological effects. While it is true that we are polluting our environment with increasingly unhealthy and unregulated dosages of electromagnetic radiation, at lower doses EMF has been shown to yield positive health effects—as SCENAR demonstrates.

HEALING

One well-researched area of EMF therapy involves how the body heals. In the 1800s, German physiologist Emil Du Bois-Reymond first reported that EMF exposure can stimulate the wound-healing process. When cells are wounded (such as when you cut yourself), the damaged cells start to leak and generate an injury current that triggers the body's natural healing processes. As Josef Penninger and Min Zhao have more recently elaborated, the injury currents direct the migration of cells into the wounded area to accelerate healing. It is now widely accepted that the flow of energy (literally, electrical charges) in your body is central to how the healing process works. Exposure to electromagnetic radiation can induce and affect these currents.

If the body can heal itself and if certain creatures can regenerate entire limbs (such as salamanders and many lizards, which can regrow their tails, complete with all its nerves, muscles, bones, and other tissue), then why are humans unable to regenerate body parts? This was the question investigated by Dr. Robert Becker, who maintained a laboratory at the Veterans Administration Hospital in Syracuse, New York.

Salamanders are able to regenerate their limbs and tails.

Capitalizing on the knowledge of injury currents, Becker attempted to stimulate the body's natural healing processes by using electrodes attached to the body to generate electric currents. This research showed Dr. Becker how sensitive cells are to even very small electric currents, and thus the potential danger to the public resulting from exposure to EMF from power lines. That EMF exposure can be both damaging and therapeutic is the theme of Becker's 1990 book, *Cross Currents*, in which he describes the known science covering both of these aspects. Becker was particularly interested in regeneration of body parts.

WIRELESS THERAPY

Dr. C. A. L. Bassett was one of the researchers in Becker's laboratory. When he later established his own laboratory in the Orthopedic Surgery Department at Columbia University, Bassett began to investigate the healing of bones by directing EMF pulses at fractures. He reported successful results, healing nonunion fractures (broken bones that had not healed naturally). Subsequent research has also found the ability of EMF exposure to stimulate regeneration in other tissues.[2]

Moreover, Bassett discovered a way to deliver therapeutic electric currents without electrodes and without having to make physical contact with the skin. He generated induced currents from electromagnetic fields in much the same way power lines generate electro-pollution currents in the environment and similar to how lightning burned the golfers (discussed in chapter 2) without actually making contact with their skin or clothes.

This wireless (and therefore, noninvasive) delivery mechanism opened up a wider range of potential EMF therapies. Indeed, Bassett and his colleagues developed a device for electromagnetic stimulation of bone healing that was among the earliest approved for use by the public. Because there are many parameters one can adjust in a complex electric signal, Bassett was optimistic about the possibility of eventually tailoring the pulsed EMF signal so that it could be effective for other medical conditions.[3]

PAIN MANAGEMENT

Another well-explored application of EMF as a therapy is in the area of pain management. One of the applications of the Russian SCENAR device, for example, is for pain relief in Olympic athletes. The FDA-approved *TENS* (Transcutaneous Electric Nerve Stimulation) machines are among the most common forms of electroanalgesia (electrical, rather than chemical, pain relief), delivering electrical current to reduce pain, particularly in those suffering from acute, chronic inflammatory conditions. Multiple separate studies have demonstrated the ability of specific types of electric fields to be an effective treatment for pain in sufferers of arthritis and fibromyalgia.[4]

Other research has shown that EMF exposure can reduce pain in sufferers of carpal tunnel syndrome over both the immediate and long term. Magnetic fields have also been found to relieve postoperative pain and reduce consumption of narcotics immediately following breast reduction surgery.[5] Similarly, sufferers of plantar fasciitis (a condition on the foot that can cause a significant amount of pain) who wore a device at night that generates a radio frequency (RF) field reported reduced pain and less need for medication.[6]

CARDIAC BYPASS

Research that Dr. Goodman and I worked on regarding the cellular stress-response stimulated by EMF (discussed in chapter 4) is the basis for a solution to a problem that frequently arises during cardiac bypass surgery.[7] In this type of surgery, the heart is immobilized, but blood continues to circulate in the body during the operation with the aid of an external pump. Problems can develop after the heart has been repaired and the body's circulation is restarted. Oxygenated blood starts to flow through cardiac tissue that has just been deprived of oxygen for hours during the surgery. The reactions of the cardiac tissue with a fresh supply of oxygen are so vigorous that they result in the accumulation of free radicals and oxidative damage to the tissue.

Studies on heart function have shown that the induction of stress proteins, as measured by the presence of hsp70, can protect the tissue from this type of damage, with resulting increases in survival rates in cardiac bypass patients. Until recently, the process used to stimulate hsp70 was an increase in temperature (recall from chapter 4, hsp stands for heat shock protein, precisely because heat was the first environmental stress recognized to stimulate the synthesis of such proteins in cells).

Unfortunately, it is rather difficult to increase the temperature of a large organ, such as the heart. The process is slow, and it is difficult to heat the heart without also heating other areas of the body near the heart. But as we've noted, stress proteins can also be induced by low-frequency EMF without the need to elevate the temperature and without the need for physical contact. Electromagnetic radiation can be delivered wirelessly, meaning that such exposures are noninvasive, and such doses of EMF can be delivered repeatedly, as a booster, as may be necessary in certain patients. Thus, exposing cardiac bypass patients to EMF has been shown to be a simpler and less damaging way to control production of hsp70 and accomplish the same result.[8]

MUSCLE STIMULATION

The stress protein synthesis used to increase survival rates in cardiac bypass patients is just one example of the biological changes that can be stimulated with exposure to electromagnetic radiation. Another example can be found in older research on protein synthesis in muscles.

Muscles are composed of interlocking thick and thin protein filaments that convert energy into tension as the filaments slide together and interact. Muscle contraction is normally initiated when the nervous system sends a message (in the form of an electrical current) to the muscle and causes the release of calcium ions that result in interactions between the thick and thin protein filaments. The interactions between the filaments generate the muscle contractions.

We know that in the absence of normal stimulation by nerves, or when nerves are cut and limbs immobilized, muscular atrophy results (i.e., muscles deteriorate from disuse). We also know that different exercises at different speeds and levels of force can be used to develop specific muscles in our bodies. Since all of this occurs via electrical messaging, we know that electrical stimulation can lead to specific changes in the composition of muscles. Electrical signals tell the genes to start producing proteins, and the characteristics of the electrical signal apparently determine which muscle-protein genes are activated.

For example, studies have shown that chronic electrical stimulation can convert skeletal muscle from fast twitch (which contract and fatigue rapidly) to slow twitch (which contract slowly, with less force). When a fast-twitch muscle, normally operating at about 100 Hz, is stimulated electrically at a much slower rate of 10 Hz, after a period of several weeks each contraction is slower and resembles that of a slow-twitch muscle. At the same time, the protein composition of the muscle tissue also changes, resembling that of a slow-twitch muscle. Other studies have shown both fast-to-slow and slow-to-fast transitions in muscle, along with corresponding changes in protein composition, due to changes in the pattern of chronic electric stimulation.[9] Thus, EMF can be applied as a force to stimulate and control the development of muscles.

TARGETED GENE ACTIVATION

Electricity is a natural mechanism used by the body to stimulate and regulate body growth through the activation of genes in your DNA. Genes can be activated by stimulation with an appropriate electrical frequency either from within the body (as occurs naturally) or supplied by electrodes attached to the body (as occurs in scientific experiments and medical therapies described earlier).

These experiments on electric stimulation in the generation of muscle tissue suggest that there may be a code relating the frequency and power of EM stimulation to the activation of a particular gene or set of genes. If this proves to be the way the system works, it should be possible to selectively stimulate an individual response by choosing a specific frequency of EM radiation. In other words, *it is theoretically possible to activate different segments of DNA, and perhaps even different individual genes, by applying different EMF frequencies.* Will science and medicine one day achieve what Drs. Becker and Bassett dreamed about by using specific EMF frequencies to activate specific healing and regeneration processes in the human body? The potential medical and therapeutic applications are thrilling to consider. Recent work described by Louis Slesin, a PhD in environmental science and editor of *Microwave News,* provides some recent evidence of this potential.

Slesin identified and publicized a series of studies showing that specific radio frequencies may be able to block the growth of specific cancers. This follows Slesin's 2007 publication summarizing similar results from other researchers. The cited papers are published in reputable scientific and medical journals. While none of this work yet indicates a cure for any particular type of cancer, this work reaffirms the possibility that it may be possible to activate selective parts of our DNA using specific EMF frequencies to stimulate processes that contribute to healing.[10]

DEVICES

Though much of the science regarding the potential of EMF as a therapy is still quite young, many medical and therapeutic devices are already available that rely on it. We've already discussed the SCENAR and TENS machines. There are many other common EMF-generating medical devices designed for taking measurements (such as temperature, blood pressure, and heart rate), as well as incubators (discussed in chapter 13). You are also likely familiar with cardiac pacemakers, which provide electrical pulses to stimulate the heart at regular intervals.

Perhaps less familiar are electroporation devices that deliver vaccines using noninvasive electrical pulses designed to increase temporarily the permeability of cell membranes. Many hospitals have therapeutic devices that expose cancer cells to microwave radiation to raise their temperature until the cells die. There are also muscle and nerve stimulators reliant upon EM exposures for their efficacy, and even a transcranial magnetic stimulator that is strong enough to interfere with the transmission of signals between nerves, enabling detailed and relatively noninvasive study of brain function.

In general, it has been difficult for therapeutic devices to obtain approval for specific conditions, and many are usually marketed simply as frequency generators or stimulators. Without entering into the controversial area of the therapeutic effectiveness of such devices, it is best to remember that EMF interactions can be harmful as well as beneficial, and the science indicates that there is a wide variation in individual sensitivities to the same exposures.

FUTURE POSSIBILITIES

There are many other areas where EMF holds therapeutic promise, beyond those already discussed in this chapter. EMF exposure has demonstrated therapeutic potential in neurodegenerative diseases, improving motor skills in Parkinson's patients.[11] Specific doses of EMF have demonstrated the short-term ability to reduce auditory hallucinations in schizophrenics.[12] EM therapy has improved rehabilitation of upper limbs and grip strength in stroke patients.[13] Electromagnetic radiation has also shown early promise in treating tinnitus, the condition in which sufferers hear a persistent ringing in the ear.[14] Other researchers found that specific doses of RF radiation aided in the healing of recalcitrant ulcers.[15] Beneficial results have even been reported in those with ADHD, who experience increased attention spans within 10 minutes after exposure to specific therapeutic doses.

All living things are complex bioelectrochemical beings, and the human body both reacts to and emits electromagnetic radiation. EMF is not only an unavoidable byproduct of modern society, it is also a critical force supporting all life as we know it. Despite the promise suggested by the above studies, there is clearly much more to learn about potential EMF therapies. Still, I believe these early results do support the notion that human-made non-ionizing electromagnetic radiation holds significant potential as a medical therapy across a wide range of human systems and ailments.

I would also underscore that the extremely low levels of EMF exposure at which many of these therapeutic outcomes are reported reinforce the fact that human-made non-ionizing EMF affects human bodies at levels that are much lower than those that are currently accepted as safe by regulatory agencies and industry. The early positive results for EMF as a therapy described here provide additional support for the argument in favor of strengthening EMF exposure safety standards.

Chapter 15

THE NEXT STEP

The fact that humans cannot see, touch, taste, or smell EMR significantly hampers people's ability to consider the risks of being exposed to it. This was a significant factor in the ready acceptance of X-ray technology when it was first introduced, with many ultimately paying the price in premature death.

This is also why accurate information about EMR is so essential. The studies presented in this book give clear evidence of significant biological effects of EMR on all life—humans, animals, plants.

These dangers are such that one would reasonably assume that action has been taken to address the situation. It has not. The primary reason for inaction is the meme repeatedly cited by the wireless industry: there is NO CONCLUSIVE PROOF of harm.

Significantly, you will note, the wireless industry does not deny that validated scientific evidence exists that many negative biological and health effects can result from exposure to the radiation emitted by their products. They focus, instead, on the lack of conclusive proof of damage to people's health.

Such an approach mandates that both industry and the general public wait until incontrovertible evidence exists that EMR damages human health. This serves only to delay the modification of industry practices to a much later point in time when a far greater number of people will have been affected by disease from these exposures.

The accumulated evidence regarding dangers of EMF exposure are clear, just as was true for tobacco, lead, asbestos, DDT, PCBs, CFCs, X-ray radiation—heck, even something as simple and obvious as getting doctors to wash their hands before operating on patients was an uphill battle against the lack of conclusive proof of harm. In each of these cases, society delayed the implementation of corrective measures until more evidence had accumulated. In the meantime, vast numbers of people were irreversibly harmed.

Today, one need only look around and see the glaciers and ice caps disappearing, the southwest United States on fire, periods of extreme temperatures (both hot and cold) becoming increasingly common, and some of our largest and most populated cities under direct threat from rising sea levels. Scientific knowledge of climate change (formerly known as global warming) has existed for almost 40 years, during which time the evidence of damage has accumulated. And still many challenge—even openly ridicule—the existence of climate change.

This ridicule is not designed to inform; rather, it is designed to mislead and delay.

This is not how we live our lives. This is not how we teach our children to live their lives. If you cross the street when the sign says Don't Walk, will you be run over and die? Probably not. But you stand a notably higher chance of dying than if you wait until the sign changes to Walk. This is why you generally don't cross the street, against the sign, unless you're in an important rush.

This is but one simple example of how we evaluate and respond to risks in our daily lives. You might use the expression "better safe than sorry" to explain it to your children.

That's how industry and regulators should respond to public health threats, such as those demonstrated with EMF exposures. Doing so involves adopting a different approach to regulations. Instead of waiting for conclusive proof that a threat exists, authorities should recognize that it is far smarter to be better safe than sorry by implementing the Precautionary Principle in a reassessment of current safety standards for EMF exposures.

As stated in the Rio Declaration, "where there are threats of serious or irreversible damage, lack of full scientific certainty shall not be used as a reason for postponing cost-effective measures to prevent environmental degradation."[1] It is vital that we act and that we do so now.

Recall, this takes place in a context in which EMF exposures continue to increase. Until a couple of hundred years ago, the only electromagnetic energy to which humans were exposed came from cosmic radiation (primarily the sun), the earth's own magnetic field, and from environmental events like lightning strikes. That's it.

Now, well over a century into the Electromagnetic Age, our world is quite different. First came the lightbulb. Then came the grid to power it. Then came electrical appliances in the home. Then came the radio. Then came radar. Then came television. Then came cordless phones. Then came microwave ovens. Then came cell phones. Then came WiFi. Then came Bluetooth devices. Then came smart meters. And so on and so on.

With each new powerful application of AC power and each new innovation in wireless communication, the human exposures continue to increase—in strength of individual exposures, in concurrency of multiple exposures, and in cumulative time exposed. We have known and been warned about this danger for years. In October 1999, the David Suzuki Foundation issued the following statement: "People in modern cities are exposed to levels of EMR millions of times higher than, and fundamentally different from, the natural background radiation. The dramatic change in our electromagnetic environment has serious implications for human health."[2] We did not evolve as biological organisms to live in this type of world, and we continue to alter the electromagnetic profile of the planet at a rate faster than life can evolve to adapt and cope with these changes. The result is genetic mutation, biological dysfunction, and disease. This trend has become increasingly clear as these unnatural sources of EMF continue to increase. With the number of devices growing apace, the EMR exposure is increasing at an enormously rapid—even exponential—rate.

In this context, with this much evidence, a wait-and-see approach makes absolutely no sense from any perspective other than the bottom-line profit of those companies that trade in EMF-emitting technologies. Instead, we must implement precautionary measures immediately at several levels: individually, within our families, and within our communities.

Chapter 12 presents the precautionary actions possible at the individual and family level. Those actions revolve around two basic rules:

1. Minimize your use of EMF-generating technologies; and
2. Maximize your distance from those EMF-generating technologies when in use.

At the community level the issues are different. One example involves the first parent who contacted me by e-mail. His children were attending a school that was planning to install a WiFi system. He wrote:

> I desperately need help. I feel . . . that continuous exposure of adolescents to RF [radio frequency] for a period of six hours a day, five days a week, for nine months out of a year, for at least three years has likely NOT been proven safe by any study. I do not know what to ask or how to frame an argument against this. I hope to challenge the representative and somehow gain the ear of a concerned decision maker to save our kids.

I responded to him with relevant information (such as some of what's been included in this book). Armed with knowledge, he crafted and distributed leaflets to inform other parents of these risks. He organized meetings and continuously worked to inform anyone who had a stake in the school's WiFi project. After several months of these activities, which catalyzed his community, *the routers in the school were removed.*

This is an outstanding and encouraging outcome and an excellent example of what can be done at the community level. One person, with no formal training in EMF science or industrial engineering, was able to obtain and package enough relevant information, to convince others, and effect real change. Using available, well-documented information, this parent's actions led to the removal of a school-wide WiFi network, creating an educational environment in which his child and the other children would avoid thousands of hours of needless exposure to microwave radiation. And this was achieved without any need for the students to forsake the Internet and its benefits, since alternative, safe systems (Ethernet cables) are available.

Having access to accurate information and putting that information into action are the keys to success. I hope this book helps by providing you with the information you need to begin creating a safer environment for yourself, your family, and your community. The public health threat resulting from EMF exposure is real, and it is well within your power to greatly reduce the risks faced by you and your loved ones.

ACKNOWLEDGMENTS

For years now, this book has been a dream of mine. Its transformation from dream to reality was made possible only through invaluable support from my family and colleagues.

I'll start with my family who played a unique, multifaceted role. My wife, Marion, is a psychologist who specializes in learning disabilities. She has long written for the general public, and just published a book on autism titled *Spectacular Bond*. Since she believes that an informed public is essential to real progress on any significant issue, she kept insisting that the results of my work have clear implications for public health and that I should move from the ivory tower of academia to the "real world." Gradually, her message took hold, and I knew that this book had to be written.

My son, Jonathan, is a polymath whose array of talents steadily amazes me. He himself is an outstanding writer. He loves finding new and interesting ideas and then getting those ideas out to the public. This led us to numerous discussions in which he steadily helped me craft different versions of the outline until we reached one that captured the message. He has been integral to the project every step of the way, coming up with solutions for every issue that came up—including introducing me to my literary manager, Peter Miller, and via Peter to my publisher, Dan Simon of Seven Stories Press.

My younger son, R, is a technologist. He is young enough to have grown up with a computer in the home as far back as he can remember, and he learned his first programming language in kindergarten. In keeping with the family tradition, R is also a writer. As I began to discuss the content of the book with him, he repeatedly helped me to reshape the ideas in ways that I had not envisioned. His suggestions had a major influence on the final product, and I am greatly indebted to him for his important role in this effort, as well as for his original diagram of the EM spectrum that shows how pervasive EMR is in our environment today.

Peter Miller, my literary manager, is an indefatigable individual with a never-ending supply of energy and enthusiasm. It is only through his support that this book came to fruition.

Dan Simon of Seven Stories has been outstanding in supporting this work as well, and he has been involved in the process at every stage. My editor at Seven Stories, Crystal Yakacki, and Elizabeth DeLong, who is in charge of production, have been a pleasure to work with, and their guidance has played a key role in the final expression that the ideas have taken.

I would also like to thank many professional colleagues for their helpful discussions over the years and especially for their suggestions in writing this book. Among these, I am pleased to mention Frank Barnes, Carl Blackman, David Carpenter, Kerry Crofton, Devra Davis, Anne Louise Gittelman, Magda Havas, Henry Lai, Blake Levitt, Abe Liboff, Bruce McLeod, Lloyd Morgan, Joel Moskowitz, Ray Neutra, Sharon Noble, Camilla Rees, Cindy Sage, Betty Sisken, and Howard Wachtel.

While focused on the writing, I feel I must thank Paul Brodeur, one of the great science reporters of our time, who was ever alert to the attempts to push new scientific discoveries ahead of our understanding the full risks associated with their development and use. Brodeur was one of the first to publicize the EMR issue, writing a three-part series in the *New Yorker* in 1989, detailing the studies showing increases in cancer associated with exposure to power lines, video display terminals, and radar microwaves. Among his many other accomplishments is the work he did in 1986 to draw attention to the depletion of the ozone layer from chlorofluorocarbon gases in aerosol sprays, and in 1989 he stressed the importance of Rachel Carson's message about the overuse of dangerous pesticides in her trailblazing book *Silent Spring*. It is unfortunate that his professional papers were not archived by the New York Public Library as originally planned.

And on behalf of those of us who have followed the EMR issue regularly for many years, let me add a special thanks to Louis Slesin, the founder and editor of *Microwave News*. This publication has been a consistent and reliable reporter on both the science and politics of EMR, and as far as we can tell, Slesin does this almost single-handedly.

Scientific research is a cooperative enterprise, and I have had the good fortune to have collaborated over several decades with many gifted colleagues. Three of the most central have been Dr. John S. Britten, Dr. Lily M. Soo, and Professor Reba M. Goodman. A pioneer in showing that EMR could activate DNA and stimulate protein synthesis in cells, Goodman was especially influential. Our discussions started many years ago and our collaboration resulted in more than 40 scientific papers along with many more abstracts of presentations at conferences and meetings. Without her involvement, none of this would have come to be. I am also grateful to Professor Andrew Marks, chairman of the Department of Physiology and Cellular Biophysics at Columbia University for his unflagging support of my work.

I shall conclude with a final note of thanks to another member of my family—my nephew Matthew Robison. While I was still searching for a title, Matt suggested "overpowered" at a family get-together. It clicked!

If I have forgotten anyone, please forgive me. It could very well be due to the effect of many years' exposure to electromagnetic radiation.☺

NOTES

CHAPTER 1: AN UNLIKELY ACTIVIST

1. Rakefet Czerninski, Avi Zini, and Harold D. Sgan-Cohen, "Risk of Parotid Malignant Tumors in Israel (1970-2006)," Epidemiology 22, no. 1 (2011): 130-131, doi: 10.1097/EDE.0b013e3181feb9f0.
2. L. Hardell, M. Carlberg, and K. Mild, Eur J Cancer Prev. 2002;159:277-283.
3. Anke Huss et al., "Residence Near Power Lines and Mortality From Neurodegenerative Diseases: Longitudinal Study of the Swiss Population," *American Journal of Epidemiology* 169, no. 2 (2008): 167-175, doi: 10.1093/aje/kwn297, http://aje.oxfordjournals.org/content/169/2/167.abstract
4. Sainudeen Sahib S., "Electromagnetic Radiation (EMR) Clashes with Honey Bees," *International Journal of Environmental Sciences* 1, no. 5 (2011), http://www.ipublishing.co.in/jesvol1no12010/EIJES2044.pdf.
5. Melinda Wenner, "Cellphone Games," *The Walrus*, September 2008, http://thewalrus.ca/cellphone-games/?ref=2008.09-health-cellphone-brain-tumour-melinda-wenner&page=.

CHAPTER 2: ELECTROMAGNETIC FIELDS

1. Vilhjalmur Rafnsson et al., "Risk of Breast Cancer in Female Flight Attendants: A Population-Based Study (Iceland)," *Cancer Causes and Control* 12, no: 2 (2001): 95-101, doi: 10.1023/A:1008983416836.
2. Yu A. Kholodov, *The Effect of Electromagnetic and Magnetic Fields on the Central Nervous System* (Moscow: NASA, 1966), http://archive.org/details/nasa_techdoc_19670022404.

CHAPTER 3: THE ELECTROMAGNETIC AGE

1. Levitt, 267.
2. *EPA: EMF In Your Environment, Magnetic Field Measurements of Everyday Electrical Devices,* 1992, 24.
3. Levitt, 336.
4. Levitt, 251.
5. "EMF Home Protection—Protect Your House from EMF," Safe Space Protection, http://www.safespaceprotection.com/electrostress-from-home-appliances.aspx.
6. "EMF In Your Environment, Magnetic Field Measurements of Everyday Electrical Devices," U.S. Environmental Protection Agency, December 1992, October 16, 2001, http://www.bls.gov/cpi/cpimwo.htm.
7. Code of Federal Regulations, U.S. Food and Drug Administration, title 21, pt. 1030, http://www.accessdata.fda.gov/scripts/cdrh/cfdocs/cfcfr/CFRSearch.cfm?CFRPart=1030.
8. Peter Leo, "Cell Phone Statistics That May Surprise You," *Pittsburgh Post-Gazette*, March 16, 2006, http://old.post-gazette.com/pg/06075/671034-294.stm.
9. Lee Rainie, "Internet, Broadband, and Cell Phone Statistics," Pew Research Center, January 5, 2010, http://pewinternet.org/~/media/Files/Reports/2010/PIP_December09_stats.pdf.
10. "ICT Data and Statistics (IDS)," International Telecommunication Union, http://www.itu.int/ITU-D/ict/statistics/ict/index.html.
11. Victor H., "Weird Statistics: More Indians Have Mobile Phones Than Have Toilets,"

PhoneArena.com, March 14, 2012, http://www.phonearena.com/news/Weird-statistics-more-Indians-have-mobile-phones-than-have-toilets_id28054.
12. http://mobithinking.com/mobile-marketing-tools/latest-mobile-stats
13. Ryan Kim, "The World's a Cell-Phone Stage/The Device is Upending Social Rules and Creating a New Culture," *San Francisco Chronicle*, February 27, 2006, http://www.sfgate.com/business/article/The-world-s-a-cell-phone-stage-The-device-is-2540683.php.
14. "Specific Absorption Rate (SAR) for Cell Phones: What it Means for You," U.S. Federal Communications Commission, last updated January 18, 2013, http://transition.fcc.gov/cgb/consumerfacts/sar.pdf.
15. "Radiation Chart," SarShield.com, http://www.sarshield.com/english/radiationchart-apple.htm.
16. http://www.mmfai.org/public/sarpdf/eng/3GS_SAR.pdf
17. "Interesting Cell Phone Statistics," SteelintheAir.com, April 20, 2005, http://www.steelintheair.com/Blog/2005/04/interesting-cell-phone-statistics.html.
18. http://www.antennasearch.com/.

CHAPTER 4: EMF DAMAMGES DNA

1. Robert Wright, "Molecular Biologists Watson and Crick," *Time Magazine*, March 29, 1999, http://www.time.com/time/magazine/article/0,9171,990626,00.html. (footnotes at end of chapter)
2. *New York Times* Oct. 1, 2002
3. Makiko Tanaka, Benjamin Elias, and Jacqueline K. Barton, "DNA-Mediated Electron Transfer in Naphthalene-Modified Oligonucleotides," *The Journal of Organic Chemistry* 75, no. 8 (2010): 2423-2428, doi: 10.1021/jo1000862.
4. H. Lai and N. P. Singh, "Magnetic-field-induced DNA Strand Breaks in Brain Cells of the Rat," *Environmental Health Perspectives* 112, no. 6 (2004): 687-694, http://www.ncbi.nlm.nih.gov/pubmed/15121512.
5. Ibid.
6. H. W. Ruediger, "Genotoxic effects of radiofrequency electromagnetic fields," *Pathophysiology* 16, no. 2-3 (2009): 67-69, doi: doi10.1016/j.pathophys.2009.02.002.
7. R. Goodman and M. Blank, (1998) *Magnetic Field Induces Expression of hsp70. Cell Stress and Chaperones* 3:79–88.
8. Di Carlo et al., "Chronic Electromagnetic Field Exposure Decreases HSP70 Levels and Lowers Cytoprotection," *Journal of Cellular Biochemistry* 84, no. 3 (2002): 447-454, http://www.ncbi.nlm.nih.gov/pubmed/11813250.
9. H. Lin, M. Blank, M. Jin, H. Lam, and R. Goodman, "Electromagnetic field stimulation of biosynthesis: changes in c-myc transcript levels during continuous and intermittent exposures," *Bioelectrochemistry and Bioenergetics* (1996) 39:215-220.
10. A.L. DiCarlo, J.M. Farrell, T.A. Litovitz, *Bioelectromagnetics*. 1998;19(8):498–500. A simple experiment to study electromagnetic field effects: protection induced by short-term exposures to 60 Hz magnetic fields.
11. Ibid.

CHAPTER 5: EMF AND CANCER

1. Judith Summers, *Soho–A History of London's Most Colorful Neighborhood* (London: Bloomsbury, 1989), 113–117.
2. http://en.wikipedia.org/wiki/1854_Broad_Street_cholera_outbreak.
3. http://www.ph.ucla.edu/epi/snow/fatherofepidemiology.html.
4. Ibid.
5. Joel M. Moskowitz, "Government Must Inform Us of Cell Phone Risk," *San Francisco Chronicle*, April 28, 2010, http://www.sfgate.com/opinion/article/Government-must-inform-us-of-cell-phone-risk-3190907.php.
6. Stephon Lonn et al., "Mobile Use and the Risk of Acoustic Neuroma," *Epidemiology* 15, no. 6 (2004): 653–659, doi: 10.1097/01.ede.0000142519.00772.bf.
7. Ibid.

8. ". . . after even just 1 or more years of use there is a 5.2-fold elevated risk in children who begin use of mobile phones before the age of 20 years, whereas for all ages the odds ratio was 1.4" Cindy Sage and David O. Carpenter, "Public Health Implications of Wireless Technologies," *Pathophysiology* 16, no. 2-3 (2009): 233–246, doi: 10.1016/j.pathophys.2009.01.011.
9. Sadetski Siegal et al., "Cellular Phone Use and Risk of Benign and Malignant Parotid Gland Tumors—A Nationwide Case-Control Study," *American Journal of Epidemiology* 167, no. 4 (2007): 457–467, http://www.ncbi.nlm.nih.gov/pubmed/18063591.
10. "Cell Phone-Cancer Link Found by Tel Aviv University Scientist," Tel Aviv University, February 14, 2008, http://www.aftau.org/site/News2?page=NewsArticle&id=6425.
11. Sadetski et al., *Pathophysiology*, 2009.
12. "Cell Phone-Cancer Link Found by Tel Aviv University Scientist," February 2008.
13. Rakefet Czerninski, Avi Zini, and Harold D. Sgan-Cohen, "Risk of Parotid Malignant Tumors in Israel (1970-2006)," *Epidemiology* 22, no. 1 (2011): 130-131, doi: 10.1097/EDE.0b013e3181feb9f0.
14. "Cellular Phone Towers," *The American Cancer Society*, last updated January 31, 2013, http://www.cancer.org/cancer/cancercauses/othercarcinogens/athome/cellular-phone-towers.
15. "Orange to Remove Mobile Mast From 'Tower of Doom,' Where Cancer Rate Has Soared," *London Evening Standard*, August 6, 2007, http://www.standard.co.uk/news/orange-to-remove-mobile-mast-from-tower-of-doom-where-cancer-rate-has-soared-7299925.html.
16. I. Yakymenko et al., "Long-Term Exposure to Microwave Radiation Provokes Cancer Growth: Evidence From Radars and Mobile Communication Systems," *Experimental Oncology* 33, no. 2 (2011): 62–70, http://www.wirelesswatchblog.org/wp-content/uploads/2011/06/Yakymenko_cancer_MW2011.pdf.
17. Eger H, Hagen K, Lucas B, et al., Einfluss der räumlichen Nähe von Mobilfunksendeanlagen auf die Krebsinzidenz, Umwelt-Medizin-Gesellschaft 2004; 17: 273–356.
18. Adilza Dode et al., "Mortality by Neoplasia and Cellular Telephone Base Stations in the Belo Horizonte Municipality," *Science of the Total Environment* (2011), doi: 10.1016/j.scitotenv.2011.05.051.
19. Neil J. Cherry, "Childhood Cancer in the Vicinity of the Sutro Tower, San Francisco," Lincoln University, September 19, 2002, http://researcharchive.lincoln.ac.nz/dspace/handle/10182/3969. http://researcharchive.lincoln.ac.nz/dspace/handle/10182/3969.
20. Orjan Hallberg and Olle Johansson, "Melanoma Incidence and Frequency Modulation (FM) Broadcasting," *Archives of Environmental Health* 57, no. 1 (2002): 32-40, doi: 10.1080/00039890209602914.
21. Orjan Hallberg and Olle Johansson, "FM Broadcasting Exposure Time and Malignant Melanoma Incidence," *Electromagnetic Biology and Medicine* 24 (2005): 1-8, doi: 10.11081/JBC-200054260.
22. Orjan Hallberg, "A Theory and Model to Explain the Skin Melanoma Epidemic," *Melanoma Research* 16, no. 2 (2006): 115–118, http://www.ncbi.nlm.nih.gov/pubmed/16567966.
23. Orjan Hallberg, "A Reduced Repair Efficiency Can Explain Increasing Melanoma Rates," *European Journal of Cancer* 17, no. 2 (2008): 147-152, doi: 10.1097/CEJ.0b013e3282b6fceb.
24. Orjan Hallberg, "Exclusive: Radio, TV Towers Linked to Increased Risk of Melanoma," Foodconsumer.org, December 9, 2007, http://foodconsumer.org/7777/8888/must-read-news/120907442007_Exclusive_report_Radio_TV_towers_linked_to_increased_risk_of_melanoma.shtm.
25. "Placement of Utility Distribution Lines Underground," Virginia Corporation Commission (2005), http://www.scc.virginia.gov/comm/reports/report_hjr153.pdf.
26. BMJ 1961; 26: 981–988.
27. Nancy Wertheimer and Ed Leeper, "Electrical Wiring Configurations and Childhood Cancer," *American Journal of Epidemiology* 109, no. 3 (1979): 273-284, http://aje.oxfordjournals.org/content/109/3/273.abstract.
28. "Electromagnetic Fields and Public Health," World Health Organization, October 2001, https://apps.who.int/inf-fs/en/fact263.html.
29. "IARC Classifies Radiofrequency Electromagnetic Fields as Possibly Carcinogenic to

Humans," World Health Organization, May 2011, http://www.iarc.fr/en/media-centre/pr/2011/pdfs/pr208_E.pdf.

CHAPTER 6: OTHER HEALTH EFFECTS OF EMF

1. Judith A. Westman et al., "Low Cancer Incidence Rates in Ohio Amish," *Cancer Causes Control* (2009), doi: 10.1007/s10552-009-9435-7.
2. "How Dirty Electricty Causes Diseases," Mercola.com, August 31, 2010, http://articles.mercola.com/sites/articles/archive/2010/08/31/how-dirty-electricity-causes-diseases.aspx.
3. Anke Huss et al., "Residence Near Power Lines and Mortality From Neurodegenerative Diseases: Longitudinal Study of the Swiss Population," *American Journal of Epidemiology* 169, no. 2 (2008): 167-175, doi: 10.1093/ajekwn297.
4. Eugene Sobel et al., "Occupations with Exposure to Electromagnetic Fields: A Possible Risk Factor for Alzheimer's Disease," *American Journal of Epidemiology* 142, no. 5 (1995): 515–524, doi: http://aje.oxfordjournals.org/content/142/5/515.short.
5. "2011 Alzheimer's Disease Facts and Figures," Alzheimer's Association, 2011, http://www.alz.org/downloads/facts_figures_2011.pdf.
6. D.A. Loomis et al., "Electrical Occupations and Neurodegenerative Disease: Analysis of U.S. Mortality Data," *Archives of Environmental Health* 53, no. 1 (1998): 71-74, http://www.ncbi.nlm.nih.gov/pubmed/9570311.
7. D.A. Savitz et al., "Magnetic Field Exposure and Neurodegenerative Disease Mortality Among Electric Utility Workers," *Epidemiology* 9, no. 4 (1998): 398-404, http://www.ncbi.nlm.nih.gov/pubmed/9647903.
8. Christoffer Johansen and Jorgen H. Olsen, "Mortality from Amyotrophic Lateral Sclerosis, Other Chronic Disorders, and Electric Shocks among Utility Workers," *American Journal of Epidemiology* 148, no. 4 (1998): 362-368, http://aje.oxfordjournals.org/content/148/4/362.short.
9. N. Hakansson et al., "Neurodegenerative Diseases in Welders and Other Workers Exposed to High Levels of Magnetic Fields," *Epidemiology* 14, no. 4 (2003): 420-426, http://www.ncbi.nlm.nih.gov/pubmed/12843765.
10. C. Qiu et al., "Occupational Exposure to Electromagnetic Fields and Risk of Alzheimer's Disease," *Epidemiology* 15, no. 6 (2004): 687-694, http://www.ncbi.nlm.nih.gov/pubmed/15475717.
11. Martin Röösli, et al, "Leukaemia, brain tumours and exposure to extremely low frequency magnetic fields: cohort study of Swiss railway employees," *Occupational and Environmental Medicine*, May 24, 2007. doi:10.1136/oem.2006.030270.
12. R.M. Park et al, "Potential Occupational Risks for Neurodegenerative Diseases," *American Journal of Industrial Medicine* 48, no. 1 (2005): 63-77, http://www.ncbi.nlm.nih.gov/pubmed/15940722.
13. Ana M Garcia, Antonio Sisternas, and Santiago Perez Hoyos, "Occupational Exposure to Extremely Low Frequency Electric and Magnetic Fields and Alzheimer Disease: A Meta-Analysis," *International Journal of Epidemiology* 37, no. 2 (2008): 329-340, doi: 10.1093/ije/dym295.
14. Brian Stein and Alasdair Philips, "The Evidence Connecting Mobile Phone EMF Exposure and Male Infertility," August 22, 2011, http://www.buergerwelle.de:8080/helma/twoday/bwnews/stories/3122/.
15. V.M, Brugh and L.I. Lipshultz, "Male Factor Infertility: Evaluation and Management," *Medical Clinics of North America* 88, no. 2 (2004): 367-385, http://www.ncbi.nlm.nih.gov/pubmed/15049583.
16. S. La Vignera et al., "Effects of the Exposure to Mobile Phones on Male Reproduction: A Review of the Literature," *Journal of Andrology* 33, no. 3 (2011): 350-356, doi: 10.2164/jandrol.111.014373.
17. A. Agarwal et al., "Effect of Cell Phone Usage on Semen Analysis in Men Attending Infertility Clinic: An Observational Study," *Fertility and Sterility* 89, no. 1 (2007): 124-128, http://www.ncbi.nlm.nih.gov/pubmed/17482179.
18. A. Agarwal et al., "Effects of Radiofrequency Electromagnetic Waves (RF-EMW) from Cellular Phones on Human Ejaculated Semen: An In Vitro Pilot Study," *Fertility and*

Sterility 92, no. 4 (2009): 1318-1325, doi: 10.1016/j.fertnstert.2008.08.022.
19. G.N. De Luliiis et al., "Mobile Phone Radiation Induces Reactive Oxygen Species Production and DNA Damage in Human Spermatozoa In Vitro," *PLoS One* 4, no. 7 (2009), doi: 10.1371/annotation/9a8a0172-3850-4059-b852-72c330769c1b.
20. Allan H. Frey, Sondra R. Feld and Barbara Frey, "Neural Function and Behavior," *Annals of the New York Academy of Sciences* 247 (1975): 433-439.
21. Leif G. Salford et al., "Permeability of the Blood-Brain Barrier Induced by 915 MHz Electromagnetic Radiation, Continuous Wave and Modulated at 8, 16, 50, and 200 Hz," *Microscopy Research and Technique* 27, no. 6 (1994): 535-542, doi: 10.1002/jemt.1070270608.
22. Nittby, et al., 2009.
23. Neil Cherry, "EMR Reduces Melatonin in Animals and People," July 26, 2000, http://www.feb.se/EMFguru/Research/emf-emr/EMR-Reduces-Melatonin.htm.
24. (In Chinese) http://www.ncbi.nlm.nih.gov/pubmed/2627835.
25. B.W. Wilson et al., "Evidence for An Effect of ELF Electromagnetic Fields on Human Pineal Gland Function," *Journal of Pineal Research* 9, no. 4 (1990): 259-269, http://www.ncbi.nlm.nih.gov/pubmed/2096195.
26. J.B. Burch et al., "Nocturnal Excretion of a Urinary Melatonin Metabolite Among Electric Utility Workers," *Scandinavian Journal of Work, Environment and Health* 24, no. 3 (1998): 183-189, http://www.ncbi.nlm.nih.gov/pubmed/9710370.
27. S. Davis, "Weak residential Magnetic Fields Affect Melatonin in Humans," *Microwave News* 17, no.6 (November/December 1997), http://microwavenews.com/news/backissues/n-d97issue.pdf.
28. J.B. Burch et al., "Melatonin Metabolite Excretion Among Cellular Telephone Users," *International Journal of Radiation Biology* 78, no. 11 (2002): 1029-1036, http://www.ncbi.nlm.nih.gov/pubmed/12456290.
29. Lucy Johnston, "Suicides Linked to Phone Masts," Express.co.uk, June 22, 2008, http://www.express.co.uk/posts/view/49330/Suicides-linked-to-phone-masts.
30. Pia K. Verkasalo et al., "Magnetic Fields of Transmission Lines and Depression," *American Journal of Epidemiology* 146, no. 12 (1997): 1037-1045, http://aje.oxfordjournals.org/content/146/12/1037.short.
31. S. Perry, L. Pearl, and R. Binns, "Power Frequency Magnetic Field; Depressive Illness and Myocardial Infarction," *Public Health* 103, no. 3 (1989): 177–180, http://www.ncbi.nlm.nih.gov/pubmed/2740472?dopt=Abstract.
32. Charles Poole et al., "Depressive Symptoms and Headaches in Relation to Proximity of Residence to an Alternating-Current Transmission Line Right-of-Way," *American Journal of Epidemiology* 137, no. 3 (1993): 318–330, http://aje.oxfordjournals.org/content/137/3/318.
33. F. Stephen Perry et al., "Environmental Power-Frequency Magnetic Fields and Suicide," *Health Physics* 41, (1981): 267–277, http://andrewamarino.com/PDFs/053-HealthPhys1981.pdf.
34. E. Van Wijngaarden et al., "Exposure to Electromagnetic Fields and Suicide Among Electric Utility Workers: A Nested Case-Control Study," *Western Journal of Medicine* 173, no. 2 (2000): 94-100, http://www.ncbi.nlm.nih.gov/pubmed/10924428.
35. I. Eliyahu et al., "Effects of Radiofrequency Radiation Emitted by Cellular Telephones on the Cognitive Functions of Humans," *Bioelectromagnetics* 27, no. 2 (2006): 119-126, http://www.ncbi.nlm.nih.gov/pubmed/16304688?dopt=Abstract; R. Luria et al., "Cognitive Effects of Radiation Emitted by Cellular Phones: The Influence of Exposure Side and Time," Bioelectromagnetics 30, no. 3 (2009): 198-204, doi: 10.1002/bem.20458; R. Hareuveny et al., "Cognitive Effects of Cellular Phones: A Possible Role of Non-Radiofrequency Radiation Factors," Bioelectromagnetics 32, no. 7 (2011): 585-588, doi: 10.1002/bem.20671.
36. "Milton Zaret, an 'Early Prophet' of Microwave Hazards, Dies at 91," *Microwave News*, June 5, 2012, http://microwavenews.com/news-center/milton-zaret-early-prophet-microwave-hazards-dies-91.
37. Allan H. Frey, "Human Auditory System Response to Modulated Electromagnetic Energy," *Journal of Applied Physiology* 17, no. 4 (1962): 689-692, http://jap.physiology.org/content/17/4/689.abstract?sid=7c073ad2-6324-4b47-94e1-124dc0a5f154.

38. J.C. Lin and Z. Wang, "Hearing of Microwave Pulses by Humans and Animals: Effects, Mechanism, and Threshholds," *Health Physics* 92, no. 6 (2007): 621–628, http://www.ncbi.nlm.nih.gov/pubmed/17495664; CHOU CK. (2007); Thirty-five years in bioelectromagnetics research. Bioelectromagnetics. 28, 3–15, http://ieeexplore.ieee.org/xpl/login.jsp?tp=&arnumber=1129245&url=http%3A%2F%2Fieeexplore.ieee.org%2Fxpls%2Fabs_all.jsp%3Farnumber%3D1129245; http://onlinelibrary.wiley.com/doi/10.1111/j.1749-6632.1975.tb35996.x/abstract; http://ieeexplore.ieee.org/xpl/login.jsp?tp=&arnumber=1455852&url=http%3A%2F%2Fieeexplore.ieee.org%2Fxpls%2Fabs_all.jsp%3Farnumber%3D1455852; http://www.ncbi.nlm.nih.gov/pubmed/1046077; http://ieeexplore.ieee.org/xpl/login.jsp?tp=&arnumber=979546&url=http%3A%2F%2Fieeexplore.ieee.org%2Fxpls%2Fabs_all.jsp%3Farnumber%3D979546; all in all, this is a good summary of links on this subject: http://www.worldcat.org/profiles/brainprivacy/lists/2914226.
39. http://emf.mercola.com/sites/emf/emf-dangers.aspx.

CHAPTER 7: THE NONHUMAN IMPACT OF EMF

1. Oskars Magone, "Latvia Sells Ghost Town," *The Baltic Times*, February 8, 2010, http://www.baltictimes.com/news/articles/24312/.
2. http://preview.wolframalpha.com/entities/cities/skrunda,_latvia/81/4n/hl/.
3. "Skrunda-1 –The Ghost Town," Way2Latvia.com, http://www.way2latvia.com/en/travel-latvia/skrunda-1.
4. "Latvian Electromagnetic Pollution," BureaudeEtudes.org, http://bureaudetudes.org/wp-content/uploads/2010/01/LatvianEMpollution.pdf.
5. T. Kalnins, R. Krizbergs, and A. Romancuks, "Measurement of the Intensity of Electromagnetic Radiation from the Skrunda Radio Location Station, Latvia," *The Science of The Total Environment* 180, no. 1 (1996): 51–56, http://www.sciencedirect.com/science/article/pii/0048969795049193.
6. I. Magone, "The Effect of Electromagnetic Radiation from Skrunda Radio Location Station on Spirodela polyrhiza (L.) Schleiden Cultures," *The Science of The Total Environment* 180, no. 1 (1996): 75–80, http://www.sciencedirect.com/science/article/pii/0048969795049223.
7. Valdis Balodis et al., "Does the Skrunda Radio Location Station Diminish the Radial Growth of Pine Trees?" *The Science of the Total Environment* 180, no. 1 (1996): 57–64, http://www.sciencedirect.com/science/article/pii/0048969795049207.
8. Turs Selga and Maija Selga, "Response of Pinus sylvestris L. Needles to Electromagnetic Fields. Cytological and Ultrastructural Aspects," *The Science of the Total Environment* 180, (1996):65–73, http://media.withtank.com/20465b47fd/response_of_pinus_sylvestris_l._needles_to_em_fields_2.pdf.
9. Z. Balode, "Assessment of Radio-Frequency Electromagnetic Radiation by the Microucleus Test in Bovine Peripheral Erythrocytes," *The Science of the Total Environment* 180, no. 1 (1996): 81-85, http://www.ncbi.nlm.nih.gov/pubmed/8717319.
10. Liepa, V. and Balodis V., "Monitoring of bird breeding near a powerful radar station. Baltic Birds," Conference on the study and conservation of birds of the Baltic region. *Vilnius*: 39. 1993.
11. T.S. Collett and J. Baron, "Biological Compasses and the Coordinate Frame of Landmark Memories in Honeybees," *Nature* 368, (1994): 137–140, doi: 10.1038/368137a0.
12. Ilia A. Solov'yov, Henrik Mouritsen, and Klaus Schulten, "Acuity of a Cryptochrome and Vision Based Magnetoreception System in Birds," *Biophysical Journal* 99, no. 1 (2010): 40–49; Ilia A. Solov'yov, Klaus Schulten, and Walter Greiner, "How Bireds and Other Animals Orient in the Earth Magnetic Field," *Physik Journal* 9, (2010) 23–28, http://www.ks.uiuc.edu/Publications/Papers/paper.cgi?tbcode=SOLO2010A; Wolfgang Wiltschko et al., "Bird Navigation: What Type of Information Does the Magnetite-Based Receptor Provide?" *Proceedings of the Royal Society: Biological Sciences* 273, no. 1603 (2006): 2815–2820.
13. Thomas P. Quinn, Ronald T. Merrill, Ernest L. Brannon, "Magnetic Field Detection in Sockeye Salmon," *Journal of Experimental Zoology* 217, no. 1 (2005): 137–142, doi: 10.1002/jez.1402170114.

14. R.A. Holland et al., "Bats Use Magnetite to Detect the Earth's Magnetic Field," *PLoS One* 3, no. 2 (2008), doi: 10.1371/journal.pone.0001676.
15. J.B. Phillips and O. Sayeed, "Wavelength-Dependent Effects of Light on Magnetic Compass Orientation in Drosophila melanogaster," *Journal of Comparative Physiology* A 172, no. 3 (1993): 303–308, http://link.springer.com/article/10.1007%2FBF00216612.
16. Blakemore, 1975.
17. S. Carrubba et al., "Evidence of a Nonlinear Human Magnetic Sense," *Neuroscience* 144, no. 1 (2006): 356-367, http://www.ncbi.nlm.nih.gov/pubmed/17069982.
18. Ed Yong, "Robins Can Literally See Magnetic Fields, but Only if Their Vision is Sharp," Discovermagazine.com, July 8, 2010, http://blogs.discovermagazine.com/notrocketscience/2010/07/08/robins-can-literally-see-magnetic-fields-but-only-if-their-vision-is-sharp/#.Ue_252QcivA.
19. A. Moller et al., "Retinal Cryptochrome in a Migratory Passerine Bird: A Possible Transducer for the Avian Magnetic Compass," *Naturwissenschaften* 91, no. 12 (2004): 585–588, http://www.ncbi.nlm.nih.gov/pubmed/15551029.
20. Dominik Heyers et al., "A Visual Pathway Links Brain Structures Active During Magnetic Compass Orientation in Migratory Birds," *PLoS One* 2, no. 9 (2007), doi: 10.1371/journal.pone.0000937.
21. Robert J. Gegear et al., "Cryptochrome Mediates Light-Dependent Magnetosensitivity in Drosphila," *Nature* 454, no. 7207 (2008): 1014–1018, doi: 10.1038/nature07183.
22. J.L. Kirschvink, M.M. Walker, and C.E. Diebel, "Magnetite-Based Magnetoreception," *Current Opinion in Neurobiology* 11, no. 4 (2001): 462–467, http://www.ncbi.nlm.nih.gov/pubmed/11502393; full: http://www.gps.caltech.edu/users/jkirschvink/pdfs/COINS.pdf.
23. J.L. Kirschvink and J.L. Gould, "Biogenic Magnetite as a Basis for Magnetic Field Detection in Animals," *Biosystems* 13, no. 3 (1981): 181–201, http://www.ncbi.nlm.nih.gov/pubmed/7213948?ordinalpos=1&itool=EntrezSystem2.PEntrez.Pubmed.Pubmed_ResultsPanel.Pubmed_SingleItemSupl.Pubmed_Discovery_RA&linkpos=1&log$=relatedarticles&logdbfrom=pubmed.
24. C. Polk, "Effects of Extremely-Low-Frequency Magnetic Fields on Biological Magnetite," *Bioelectromagnetics* 15, no. 3 (1994): 261–270, http://www.ncbi.nlm.nih.gov/pubmed/8074740?ordinalpos=1&itool=EntrezSystem2.PEntrez.Pubmed.Pubmed_ResultsPanel.Pubmed_SingleItemSupl.Pubmed_Discovery_RA&linkpos=1&log$=relatedarticles&logdbfrom=pubmed.
25. Wolfgang Wiltschko et al., "Bird Navigation: What Type of Information Does the Magnetite-Based Receptor Provide?" *Proceedings of the Royal Society: Biological Sciences* 273, no. 1603 (2006): 2815–2820, http://rspb.royalsocietypublishing.org/content/273/1603/2815.short.
26. U. Munro et al., "Evidence for a Magnetite-Based Navigational "Map" in Birds," *Naturwissenschaften* 84, no. 1 (1997): 26-28, http://link.springer.com/.
27. Aronoff, A. 1949. The September migration tragedy. Linnaean News-Letter 31): 2.
28. Albert M. Manville, "Bird Strikes and Electrocutions at Power Lines, Communication Towers, and Wind Turbines: State of the Art and State of the Science –Next Steps Toward Mitigation," USDA Forest Service General Tech Report (2005), http://www.fs.fed.us/psw/publications/documents/psw_gtr191/Asilomar/pdfs/1051-1064.pdf.
29. R. L. Crawford and R. T. Engstrom, "Lights, Towers, and Avian Mortality: Where is the Science?" Proceedings of the workshop on avian mortality at communication towers, August 11, 1999, Cornell University, Ithaca, NY, http://www.towerkill.com/activism/workshop/pdf/amact.pdf.
30. R.L. Crawford and R. T. Engstrom, "Characteristics of Avian Mortality at a North Florida Television Tower: A 29-Year Study," *Journal Field Ornithology* 72, no. 3 (2001): 380–388, http://www.bioone.org/doi/abs/10.1648/0273-8570-72.3.380.
31. Charles Kemper, "A Study of Bird Mortality at a West Central Wisconsin TV Tower from 1957-1995," *The Passenger Pigeon* 58, (1996): 219–235, http://images.library.wisc.edu/EcoNatRes/EFacs/PassPigeon/ppv58no03/reference/econatres.pp58no3.ckemper.pdf.
32. Albert M. Manville, "Bird Strikes and Electrocutions at Power Lines, Communication Towers, and Wind Turbines: State of the Art and State of the Science –Next Steps Toward Mitigation," USDA Forest Service General Tech Report (2005), http://www.fs.fed.us/psw/

publications/documents/psw_gtr191/Asilomar/pdfs/1051-1064.pdf.
33. "Briefing Paper on the Need for Research into the Cumulative Impacts of Communication Towers on Migratory Birds and Other Wildlife in the United States," Division of Migratory Bird Management (DMBM) and U.S. Fish and Wildlife Service, April 2009, http://electromagnetichealth.org/pdf/CommTowerResearchNeedsPublicBriefing-2-409.pdf
34. Ibid.
35. http://www.es-uk.info/research/20090606_goldsworthy_birds_bees_emf.pdf.
36. Alfonso Balmori Martinez, "The Effects of Microwave Radiation on the Wildlife," February 2003, http://www.emrpolicy.org/litigation/case_law/beebe_hill/balmori_wildlife_study.pdf.
37. Joris Everaert and Dirk Bauwens, "A Possible Effect of Electromagnetic Radiation from Mobile Phone Base Stations on the Number of Breeding House Sparrows (Passer domesticus)," *Electromagnetic Biology and Medicine* 26, (2007): 63-72, doi: 10.1080/15368370701205693.
38. A. Balmori and O. Hallberg, "The Urban Decline of the House Sparrow (Passer domesticus): A Possible Link with Electromagnetic Radiation," *Electromagnetic Biology and Medicine* 26, no. 2 (2007): 141-151, http://www.ncbi.nlm.nih.gov/pubmed/17613041.
39. B.J. Youbicier-Simo, J.C. Lebecq, and M. Bastide, "Mortality of Chicken Embryos Exposed to EMFs from Mobile Phones," presented at the Twentieth Annual Meeting of the Bioelectromagnetics Society, St. Pete Beach, FL, http://electricwords.emfacts.com/y013240.html.
40. http://en.scientificcommons.org/20212656.
41. Kimberly J. Fernie and David M. Bird, "Effects of Electromagnetic Fields on Body Mass and Food-Intake of American Kestrels," *The Condor* 101, no. 3 (1999): 616–621, http://www.avaate.org/IMG/pdf/fernie_cernicalos.pdf.
42. Kimberly J. Fernie, David M. Bird, and Denis Petitclerc, "Effects of Electromagnetic Fields on Photophasic Circulating Melatonin Levels in American Kestrels," *Environmental Health Perspectives* 107, no. 11 (1999): 901–904, http://www.ncbi.nlm.nih.gov/pmc/articles/PMC1566687/pdf/envhper00516-0087.pdf.
43. Kimberly J. Fernie and David M. Bird, "Evidence of Oxidative Stress in American Kestrels Exposed to Electromagnetic Fields," *Environmental Research* 86, no. 2 (2001): 198–207, http://www.sciencedirect.com/science/article/pii/S001393510194263X.
44. Kimberly J. Fernie et al., "Effects of Electromagnetic Fields on the Reproductive Success of the American Kestrel," *Physiological and Biochemical Zoology* 73, no. 1 (2000): 60–65, http://www.jstor.org/discover/10.1086/316726?uid=2&uid=4&sid=21102558252707.
45. Allison Benjamin, "Fears for Crops as Shock Figures From America Show Scale of Bee Catastrophe," *The Guardian*, May 1, 2010, http://www.guardian.co.uk/environment/2010/may/02/food-fear-mystery-beehives-collapse.
46. Renee Johnson, "Honey Bee Colony Collapse Disorder," CRS Report for Congress, January 7, 2010, http://www.fas.org/sgp/crs/misc/RL33938.pdf.
47. Benjamin, *Guardian*, May 1, 2010.
48. Ibid.
49. "Why are Europe's Bees Dying?" BBC News, November 20, 2008, http://news.bbc.co.uk/2/hi/europe/7739798.stm.
50. Benjamin, *Guardian*, 2010
51. http://www.greenearthfriend.com/2009/01/colony-collapse-disorder-ccd-honeybees-dying-by-the-millions/.
52. "Why are Europe's Bees Dying?" BBC News, November 20, 2008, http://news.bbc.co.uk/2/hi/europe/7739798.stm.
53. Paul Molga, "La Mort des Abeilles Met la Planète en Danger La mort des abeilles met la planète en danger," *Les Echos*, August 20, 2007, http://www.lesechos.fr/20/08/2007/LesEchos/19985-63-ECH_la-mort-des-abeilles-met-la-planete-en-danger.htm.
54. J.L. Kirschvink and J.L. Gould, "Biogenic Magnetite as a Basis for Magnetic Field Detection in Animals," *Biosystems* 13, no. 3 (1981): 181–201, http://www.ncbi.nlm.nih.gov/pubmed/7213948/.
55. Andrew Goldsworthy, "The Birds, the Bees and Electromagnetic Pollution," MastSanity.org, May 2009, http://www.mastsanity.org/health-52/research/269-the-birds-the-bees-and-

electromagnetic-pollution-by-dr-andrew-goldsworthy-may-2009.html.

56. Stefan Kimmel et al., "Electromagnetic Radiation: Influences on Honeybees (Apis mellifera)," http://www.hese-project.org/hese-uk/en/papers/kimmel_iaas_2007.pdf.
57. S. Sainudeen Sahib, "Electromagnetic Radiation (EMR) Clashes with Honey Bees," *International Journal of Environmental Sciences* 1, no. 5 (2011), http://www.ipublishing.co.in/jesvol1no12010/EIJES2044.pdf.
58. V.P. Sharma and N.R. Kumar, "Changes in Honeybee Behaviour and Biology Under the Influence of Cellphone Radiations," *Current Science* 98, no. 10 (2010): 1376–1378, http://www.scribd.com/doc/32303006/Changes-in-Honeybee-Behaviour-and-Biology-Under-the-Influence-of-Cellphone-Radiations.
59. Daniel Favre, "Mobile Phone-Induced Honeybee Worker Piping," *Apidologie* 42, no. 3 (2011): 270-279, http://link.springer.com/article/10.1007%2Fs13592-011-0016-x.
60. http://www.mobiledia.com/news/90232.html.
61. Alain Vian et al., "Plants Respond to GSM-Like Radiation," *Plant Signaling and Behavior* 2, no. 6 (2007): 522-524, http://www.ncbi.nlm.nih.gov/pmc/articles/PMC2634357/.
62. Katie Haggerty, "Adverse Influence of Radio Frequency Background on Trembling Aspen Seedlings: Preliminary Observations," *International Journal of Forestry Research* 2010 (2010), doi: 10.1155/2010/836278.
63. David D. Reed et al., "Effects of 76 Hz Electromagnetic Fields on Forest Ecosystems in Northern Michigan: Tree Growth," *International Journal of Biometeorology* 37, no. 4 (1993): 229-234, doi: 10.1007/BF01387529.
64. Alfonso Balmori Martinez, "The Effects of Microwaves on the Trees and Other Plants," December 2003, http://www.boomaantastingen.nl/onderzoek_bomen_planten.pdf.
65. First impression Knowledge Platform on Research and trees WIFI broadcast signals; http://translate.google.com/translate?hl=en&sl=nl&tl=en&u=http://www.antennebureau.nl/actueel/nieuws/2010/eerste-indruk-kennisplatform-onderzoek-naar-bomen-en-wifi-zendsignalen; Summarized: http://www.popsci.com/technology/article/2010-11/wi-fi-radiation-killing-trees.
66. http://news.discovery.com/earth/power-lines-contaminants-pcb.html; A.P. Fews et al., "Increased Exposure to Pollutant Aerosols Under High Voltage Power Lines," *International Jounral of Radiation Biology* 75, no. 12 (1999): 1505-1521, http://www.mendeley.com/research/increased-exposure-pollutant-aerosols-under-high-voltage-power-lines/.
67. D.D. Sandu et al., "A Preliminary Study on Ultra High Frequency Electromagnetic Fields Effect on Black Locust Chlorophylls," *Acta Biologica Hungarica* 56, no. 1–2 (2005): 109–117, http://www.ncbi.nlm.nih.gov/pubmed/15813219?dopt=Abstract.
68. D. Roux et al., "High Frequency (900 MHz) Low Amplitude (5 V m-1) Electromagnetic Field: A Genuine Environmental Stimulus that Affects Transcription, Translation, Calcium and Energy Charge in Tomato," *Planta* 227, no. 4 (2008): 883–891, http://www.ncbi.nlm.nih.gov/pubmed/18026987.
69. E. Ben-Izhak Monselise, A.H. Parola, and D. Kost, "Low-Frequency Electromagnetic Fields Induce a Stress Effect Upon Higher Plants, as Evident by the Universal Stress Signal, Alanine," *Biochemical and Biophysical Research Communications* 302, no. 2 (2003): 427–434, http://www.ncbi.nlm.nih.gov/pubmed/12604366.
70. M. Tkalec, K. Malaric, and B. Pevalek-Kozlina, "Influence of 400, 900, and 1900 MHz Electromagnetic Fields on Lemna minor Growth and Peroxidase Activity," *Bioelectromagnetics* 26, no. 3 (2005): 185–193, http://www.ncbi.nlm.nih.gov/pubmed/15768427.
71. M. Tkalec, K. Malaric, and B. Pevalek-Kozlina, "Exposure to Radiofrequency Radiation Induces Oxidative Stress in Duckweed Lemna Minor L.," *Science of the Total Environment* 388, no. 1-3 (2007): 78–89, http://www.ncbi.nlm.nih.gov/pubmed/17825879.
72. H. Sahebjamei, P. Abdolmaleki, and F. Ghanati, "Effects of Magnetic Field on the Antioxidant Enzyme Activities of Suspension-Cultured Tobacco Cells," *Bioelectromagnetics* 28, no. 1 (2007): 42–47, http://www.ncbi.nlm.nih.gov/pubmed/16988990.
73. G. Soja et al., "Growth and Yield of Winter Wheat (Triticum aestivum L.) and Corn (Zea mays L.) Near a High Voltage Transmission Line," *Bioelectromagnetics* 24, no. 2 (2003): 91–102, http://www.ncbi.nlm.nih.gov/pubmed/12524675.
74. L.G. Kallinin et al., "The Influence of a Low- and High-Frequency Electromagnetic

Fields on Seeds," *Biofizika* 50, no. 2 (2005): 361–366, http://www.ncbi.nlm.nih.gov/pubmed/15856999.
75. Sharma V.P. et al, "Cell Phone Radiations Affect Early Growth of Vigna radiata (Mung Bean) Through Biochemical Alterations," *Verlag der Zeitschrift fur Naturforschung* 65, no. 1-2 (2010): 66–72, http://www.ncbi.nlm.nih.gov/pubmed/20355324.
76. H.H. Huang and S.R. Wang, "The Effects of Inverter Magnetic Fields on Early Seed Germination of Mung Beans," *Bioelectromagnetics* 29, no. 8 (2008): 649–657, doi: 10.1002/bem.20432.

CHAPTER 8: THE BUSINESS OF EMF SCIENCE

1. Harold J. Cook et al., "Early Research on the Biological Effects of Microwave Radiation: 1940-1960," *Annals of Science* 37, no. 3 (1980): 323–351, doi: 10.1080/00033798000200271.
2. http://www.magdahavas.com/pick-of-the-week-20-early-research-on-the-biological-effects-of-microwave-radiation-1940-1960/p 330.
3. http://www.zoryglaser.com/.
4. http://www.magdahavas.com/wordpress/wp-content/uploads/2010/06/Navy_Radiowave_Brief.pdf.
5. Dr. Magda Havas is in the process of digitizing Dr. Zory Glaser's work, which can be viewed at: http://www.magdahavas.com/category/from-zorys-archive/.
6. Zorach Glaser, "Bibliography of Reported Biological Phenomena ('Effects') and Clinical Manifestations Attributed to Microwave and Radio-Frequency Radiation," *Naval Medical Research Institute*, October 4, 1971, http://www.magdahavas.com/wordpress/wp-content/uploads/2010/06/Navy_Radiowave_Brief.pdf.
7. Dr. Magda Havas is in the process of digitizing Dr. Zory Glaser's work, which can be viewed at: http://www.magdahavas.com/category/from-zorys-archive/.
8. Ibid.
9. Ibid.
10. Ibid.
11. Ibid.
12. Rowland Evans and Robert Novak, "Soviet Microwaves and a Diplomat's Death," *Lawrence Journal-World*, December 14, 1986.
13. "Mobile Phone Subscribers (1989) by Country," NationMaster.com, http://www.nationmaster.com/graph/med_mob_pho_sub-media-mobile-phone-subscribers&date=1989.
14. "Biological Effects of Power Frequency Electric and Magnetic Fields—Background Paper," U.S. Congress Office of Technology Assessment, May 1989, http://ota.fas.org/reports/8905.pdf.
15. "Electromagnetic Fields and Cancer—Media and Public Attention Affect Research," *Journal of the National Cancer Institute* 83, no. 3 (1991): 164–166, doi: 10.1093.jnci.83.3.164.
16. http://ofmpub.epa.gov/eims/eimscomm.getfile?p_download_id=437194
17. Philip J. Hilts, "Study Says Electrical Fields Could be Linked to Cancer," *New York Times*, December 15, 1990, http://www.nytimes.com/1990/12/15/us/study-says-electrical-fields-could-be-linked-to-cancer.html.
18. David Hafmeister, "Background Paper on Power Line Fields and Public Health," to the American Physical Society Panel on Public Affairs, March 29, 1996, http://www.ehso.com/ehshome/emfstudy.htm.
19. "Questions and Answers About Electric and Magnetic Fields (EMFs)," U.S. Environmental Protection Agency, December 1992, http://nepis.epa.gov.
20. *Microwave News*, M/J90 and N/D90.
21. *Microwave News*, M/A95.
22. *Microwave News*, S/O95.
23. "EPA Shelves EMF-Cancer Report But Link Called Stronger Than in 1990," *Microwave News* 16, no. 1 (January/February 1996), http://microwavenews.com/sites/default/files/sites/default/files/backissues/j-f96issue.pdf.
24. "Electric and Magnetic Fields," National Institute of Environmental Health Sciences, last

reviewed May 16, 2013, http://www.niehs.nih.gov/health/topics/agents/emf/.

25. "Mobile Phone Subscribers (1989) by Country," NationMaster.com, http://www.nationmaster.com/graph/med_mob_pho_sub-media-mobile-phone-subscribers&date=1989.
26. James Martinez, "Widower on Mission Against Cellular Phones," *The Daily Gazette*, February 7, 1993.
27. Barbara Grady, "Telephone Firms Fight Cancerous Connection," *Pittsburgh Post-Gazette*, January 26, 1993.
28. Val Ellicott, "Woman's Death Fuels Phone Fears Cancer Scare Rocks Cellular Industry," *The Palm Beach Post*, February 1, 1993.
29. Anthony Ramirez, "Cellular Phone Scare Hits Stocks and Worries Companies," *Sarasota Herald-Tribune*, January 29, 1993.
30. Ibid.
31. http://articles.sun-sentinel.com/2005-10-02/features/0509290643_1_phone-radiation-cell-phone-cellular-telephone
32. Jeffrey Silva, "Motorola Memo Raises Questions About WTR Research," RCRWireless.com, March 3, 1997, http://www.rcrwireless.com/article/19970303/sub/motorola-memo-raises-questions-about-wtr-research/.
33. "George Carlo, PhD, JD" ProCon.org, last updated February 19, 2010, http://cellphones.procon.org/view.source.php?sourceID=009487.
34. Melinda Wenner, "Cellphone Games," *The Walrus*, September 2008, http://thewalrus.ca/cellphone-games/?ref=2008.09-health-cellphone-brain-tumour-melinda-wenner&page=.
35. Naomi Ishisaka, "UW Scientist Henry Lai Makes Waves in the Cell Phone Industry," *Seattle Magazine*, January 2011, http://www.seattlemag.com/article/nerd-report/nerd-report.
36. Wenner, "Cell Phone Games," September 2008.
37. "Cell Phone Safety and the Industry-Part 2," transcription of KGNU radio show *La Vida!* interview with Jerry Phillips, August 13, 2006, http://lavida.kgnu.net/lavidaradioshow.php?show_id=268.
38. Wenner, "Cell Phone Games," September 2008.
39. "Cell Phone Safety and the Industry-Part 2," transcription of KGNU radio show *La Vida!* interview with Jerry Phillips, August 13, 2006, http://lavida.kgnu.net/lavidaradioshow.php?show_id=268.
40. Ibid.
41. Ibid.
42. "Mobile Phone Health Risks," transcription of BBC 1 television show *Panorama* interview with reporter Paul Kenyon and Dr. George Carlo, May 5, 1999, http://www.reocities.com/researchtriangle/9884/orange_ltd__a_panorama_report.htm.
43. Rob Harrill, "Wake-Up Call," *Columns*, March 2005, http://www.washington.edu/alumni/columns/march05/wakeupcall02.html.
44. http://lavida.kgnu.net/lavidaradioshow.php?show_id=269.
45. J. L. Phillips, NP Singh, H Lai (2009) Electromagnetic fields and DNA damage. Pathophysiology 16:79–88.
46. "From the Field," *Microwave News* 16, no. 1 (January/February 1996), http://microwavenews.com/sites/default/files/sites/default/files/backissues/j-f96issue.pdf.
47. "Cell Phone Safety and the Industry-Part 3," transcription of KGNU radio show *La Vida!* interview with Jerry Phillips, August 20, 2006, http://lavida.kgnu.net/lavidaradioshow.php?show_id=269.
48. http://microwavenews.com/sites/default/files/sites/default/files/backissues/j-f98issue.pdf; http://microwavenews.com/sites/default/files/sites/default/files/backissues/m-a98issue.pdf
49. "Members of the Panels," *Microwave News* 14, no. 4 (July/August 1994), http://microwavenews.com/news/backissues/j-a94issue.pdf.
50. Repacholi et al., Radiat. Res. 147, 631–640, 1997.
51. Utteridge et al, Radiat. Res. 158, 357–364, 2002
52. Wenner, "Cell Phone Games," September 2008.
53. http://microwavenews.com/news/backissues/j-f98issue.pdf.
54. Joachim Schuz et al., "Cellular Telephone Use and Cancer Risk: Update of a Nationwide

Danish Cohort," *Journal of the National Cancer Institute* 98, no. 23 (2006): 1707-1713, doi: 10.1093/jcni/djj464.
55. http://www.canceractive.com/cancer-active-page-link.aspx?n=1541
56. "The Danish Cohort Study: The Politics and Economics of Bias," MicrowaveNews.com, November 3, 2011, http://microwavenews.com/DanishCohort.html.
57. Lennart Hardell and Kjell Hansson Mild, "Re: Cellular Telephones and Cancer—A Nationwide Cohort Study in Denmark," *Journal of the National Cancer Institute* 98, no. 12 (2001): 952, doi: 10.1093/jnci/93.12.952.
58. Miranda Hitti, "No Cancer Risk Seen With Cell Phones," CBSNews.com, December 5, 2006, http://www.cbsnews.com/stories/2006/12/05/health/webmd/main2232261.shtml.
59. Lauran Neergaard, "Study Disputes Cellphone-Cancer Link," USAToday.com, December 5, 2006, http://usatoday30.usatoday.com/tech/news/2006-12-05-cellphones-cancer_x.htm.
60. http://nl.newsbank.com/nl-search/we/Archives?p_product=SJ&s_site=mercurynews&p_multi=SJ&p_theme=realcities&p_action=search&p_maxdocs=200&p_topdoc=1&p_text_direct-0=115DF7C47903A078&p_field_direct-0=document_id&p_perpage=10&p_sort=YMD_date:D&s_trackval=GooglePM.
61. "Power-Line EMFs: New Focus on Alzheimer's Disease," MicrowaveNews.com, November 17, 2008, http://microwavenews.com/adroosli.html.
62. Naomi Ishisaka, "UW Scientist Henry Lai Makes Waves in the Cell Phone Industry," *Seattle Magazine*, January 2011, http://www.seattlemag.com/article/nerd-report/nerd-report.
63. Christina Turner and George J. Spilich, "Research into Smoking or Nicotine and Human Cognitive Performance: Does the Source of Funding Make a Difference?" *Addiction* 92, no. 11 (1997): 1432-1426, doi: 10.1111/j.1360-0443.1997.tb02863.x.
64. David Michaels, "It's not the Answers That Are Biased, It's the Questions," *The Washington Post*, July 15, 2008, http://www.washingtonpost.com/wp-dyn/content/article/2008/07/14/AR2008071402145.html.
65. W. Becker-Bruser, "Research in the Pharmaceutical Industry Cannot be Objective," *Zeitschrift fur Evidenz, Fortbildung und Qualitat im Gesundheitswesen* 104, no. 3 (2010): 183–189, http://www.ncbi.nlm.nih.gov/pubmed/20608245.
66. Jeffry Fawcett, "WiFi Blues," Arizona Center for Advanced Medicine, June 26, 2013, http://arizonaadvancedmedicine.com/wifi-blues-by-jeffry-fawcett-phd/.
67. Wenner, "Cell Phone Games," September 2008.
68. Anke Huss et al., "Source of Funding and Results of Studues of Health Effects of Mobile Phone Use: Systematic Review of Experimental Studies," *Environmental Health Perspectives* 115, no. 1 (2007), doi: 10.1289/ehp.9149.
69. http://walrusmagazine.com/articles/2008.09-health-cellphone-brain-tumour-melinda-wenner/.
70. http://ehp03.niehs.nih.gov/article/fetchArticle.action?articleURI=info:doi/10.1289/ehp.9149.
71. Joel M. Moskowitz, "Government Must Inform Us of Cell Phone Risk," SFGate.com, July 25, 2013, http://www.sfgate.com/cgi-bin/article.cgi?f=/c/a/2010/04/27/EDMB1D58TC.DTL#ixzz1qAghpiqI.
72. Rob Harrill, "Wake-Up Call," *Columns*, March 2005, http://www.washington.edu/alumni/columns/march05/wakeupcall02.html.
73. Stephen Adams, "No Evidence that Mobiles Cause Cancer – but Scientists Still Can't Rule it Out," *The Telegraph*, April 26, 2012, http://www.telegraph.co.uk/health/healthnews/9226041/No-evidence-that-mobiles-cause-cancer-but-scientists-still-cant-rule-it-out.html.
74. http://emrstop.net/index.php?option=comkunena&func=view&catid=5&id=144&Itemid=19.
75. Mona Nilsson, "Conflict of Interest at WHO," May 23, 2011, http://www.monanilsson.se/document/AhlbomConflictsIARCMay23.pdf.
76. Ibid.
77. Lennart Hardell et al., "Secret Ties to Industry and Conflicting Interests in Cancer Research," *American Journal of Industrial Medicine* 50, no. 3 (2007): 235-236, doi: 10.1002/ajim.20426.
78. "The Fallout From the Telecommunications Act of 1996: Unintended Consequences

and Lessons Learned," Common Cause Education Fund, May 9, 2005, http://www.commoncause.org.
79. Christopher Ketcham, "Warning: Your Cell Phone May Be Hazardous to Your Health," *GQ*, February 2010, http://www.gq.com/cars-gear/gear-and-gadgets/201002/warning-cell-phone-radiation.
80. http://www.ccst.us/projects/smart/documents/li_response.pdf.
81. http://www.ccst.us/projects/smart2/documents/letter8hirsch.pdf.
82. Maureen Dowd, "Are Cells the New Cigarettes?" *The New York Times*, June 26, 2010, http://www.nytimes.com/2010/06/27/opinion/27dowd.html?_rhttp://www.nytimes.com/2010/06/27/opinion/27dowd.html?_r=3=3&_r=0.
83. Maureen Dowd, "Are Cells the New Cigarettes?" *The New York Times*, June 26, 2010, http://www.nytimes.com/2010/06/27/opinion/27dowd.html?_rhttp://www.nytimes.com/2010/06/27/opinion/27dowd.html?_r=3=3&_r=0.
84. Ibid.
85. George L. Carlo, "The Latest Reassurance Ruse About Cell Phones and Cancer," *Science and Public Policy Institute*, http://www.powerwatch.org.uk/pdfs/20070105_conflict_carlo.pdf.
86. "Cell Phone Safety and the Industry-Part 2," transcription of KGNU radio show *La Vida!* interview with Jerry Phillips, August 13, 2006, http://lavida.kgnu.net/lavidaradioshow.php?show_id=268.
87. Jennifer Shin, "Wireless Industry Reached Profits of Almost $19 Billion in First Quarter," BQSoft.com, June 1, 2012, http://www.bqsoft.com/2012/06/wireless-industry-reached-profits-of-almost-19-billion-in-first-quarter/.
88. Ketcham, *GQ*, February 2010.
89. Chris Mooney, "The Manufacture of Uncertainty," *The American Prospect*, March 19, 2008, http://prospect.org/article/manufacture-uncertainty.

CHAPTER 9: DOUBT, FROM TOBACCO TO INTERPHONE

1. Maureen Dowd, "Are Cells the New Cigarettes?" *The New York Times*, June 26, 2010, http://www.nytimes.com/2010/06/27/opinion/27dowd.html?_rhttp://www.nytimes.com/2010/06/27/opinion/27dowd.html?_r=3=3&_r=0.
2. "Study: Cell Phones Could Be More Dangerous Than Cigarettes," FoxNews.com, March 31, 2008, http://www.foxnews.com/story/0,2933,343335,00.html.
3. Molly Wood, "The Cell Phone Industry: Big Tobacco 2.0?" CNet.com, March 8, 2005, http://www.cnet.com/4520-6033_1-5741203-1.html.
4. "Smoking and Health Proposal," TobaccoDocuments.org, 1969, http://tobaccodocuments.org/landman/332506.html.
5. David Michaels, *Doubt is Their Product: How Industry's Assault on Science Threatens Your Health*, (New York: Oxford University Press, 2008).
6. "In Old Ads, Doctors and Babies Say 'Smoke,'' *The New York Times*, http://www.nytimes.com/slideshow/2008/10/06/business/media/20081006_CigaretteAd_Slideshow_ready_3.html.
7. Ernest L. Wynder and Evarts A. Graham, "Tobacco Smoking as a Possible Etiologic Factor in Bronchiogenic Carcinoma," *Journal of American Medicine* 143, no. 4 (1950): 329–336, doi: 10.1001/jama.1950.02910390001001.
8. Gene Borio, "Tobacco Timeline: The Twentieth century 11950-1999—The Battle is Joined," Tobacco.org, http://archive.tobacco.org/resources/history/Tobacco_History20-2.html
9. http://www.tobacco.neu.edu/litigation/cases/supportdocs/frank_ad.htm.
10. "A Frank Statement to Cigarette Smokers," Tobacco.org, http://archive.tobacco.org/History/540104frank.html.
11. Lennart Hardell et al., "Secret Ties to Industry and Conflicting Interests in Cancer Research," *American Journal of Industrial Medicine* 50, no. 3 (2007): 235-236, doi: 10.1002/ajim.20426.
12. "Ragner Rylander vs CIPRET and OxyGeneve," *Prevention*, May 24, 2002, http://www.prevention.ch/ryjugementen.htm.
13. "031215 Final Rylander Judgment in Geneva," decision of the Court of Appeal, December 15, 2003, Tobacco.org, http://archive.tobacco.org/resources/documents/031215rylander.html.

14. "Inform Me: Timeline," Tobacco Technical Assistance Consortium, http://learningcenter.ttac.org/learning/comp04/04A_INF001.asp.
15. Australia, Canada, Denmark, Finland, France, Germany, Israel, Italy, Japan, New Zealand, Norway, Sweden and the UK.
16. "The Interphone Study," International Agency for Research on Cancer and World Health Organization, http://interphone.iarc.fr/.
17. Christopher Wild, "IARC Report to the Union for Internatonal Cancer Control (UICC) on Interphone Study," International Agency for the Research on Cancer and World Health Organization, October 3, 2011, http://interphone.iarc.fr/UICC_Report_Final_03102011.pdf.
18. E. Cardis et al., "The INTERPHONE Study: Design, Epidemiological Methods, and Description of the Study Population," *European Journal of Epidemiology* 22, no. 9 (2007): 647–664, doi: http://www.ncbi.nlm.nih.gov/pubmed/17636416; E. Cardis et al., "Brain Tumour Risk in Relation to Mobile Telephone Use: Results of the INTERPHONE International Case Control Study," *International Journal of Epidemiology* 39, no.3 (2010): 675–694, doi: 10.1093/ije/dyq079; E. Cardis et al., "Acoustic Neuroma Risk in Relation to Mobile Telephone Use: Results of the INTERPHONE International Case-Control Study," *Cancer Epidemiology* 35, no. 5 (2011): 453–464, doi:10.1016/j.canep.2011.05.012.
19. M. Vrijheid et al., "Validation of Short Term Recall of Mobile Phone Use for the Interphone Study," *Occupation Environment Medicine* 63, no. 4 (2006): 237–243, doi: 10.1136/oem.2004.019281; M. Vrijheid et al., "Quantifying the Impact of Selection Bias Caused by Nonparticipation in a Case Control Study of Mobile Phone Use," *Annals of Epidemiology* 19, no. 1 (2009): 33–41, doi: 10.1016/j.annepidem.2008.10.006; M. Vrijheid et al., "Recall Bias in the Assessment of Exposure to Mobile Phones," *Journal of Exposure Science and Environmental Epidemiology* 19, no. 4 (2009): 369–381, doi: 10.1038/jes.2008.27; M. Vrijheid et al., "Determinants of Mobile Phone Output Power in a Multinational Study: Implications for Exposure Assessment," *Occupation Environment Medicine* 66, no. 10 (2009): 664–671, doi: 10.1136/oem.2008.043380.
20. See the IARC Report to the Union for International Cancer Control (UICC) on the Interphone Study at http://interphone.iarc.fr/UICC_Report_Final_03102011.pdf for full listing.
21. "Interphone Study Reports on Mobile Phone Use and Brian Cancer Risk," International Agency for Research on Cancer and World Health Organization, May 17, 2010, http://www.iarc.fr/en/media-centre/pr/2010/pdfs/pr200_E.pdf.
22. Ibid.
23. "The Interphone Study-Funding," International Agency for Research on Cancer and World Health Organization, http://interphone.iarc.fr/interphone_funding.php.
24. Siddhartha Mukherjee, "Do Cell Phones Cause Brain Cancer?" *The New York Times*, April 13, 2011, http://www.nytimes.com/2011/04/17/magazine/mag-17cellphones-t.html?pagewanted=all&_r=0.
25. "Mobile Madness," *The Economist*, September 25th, 2008, http://www.economist.com/node/12295222.
26. Australia, Canada, France, Germany, Italy, Japan and New Zealand.
27. Christopher Wild, "IARC Report to the Union for Internatonal Cancer Control (UICC) on Interphone Study," International Agency for the Research on Cancer and World Health Organization, October 3, 2011, http://interphone.iarc.fr/UICC_Report_Final_03102011.pdf.
28. "Cellphones and Cancer: Watch That Thing by Your Head," *The Globe and Mail*, May 20, 2010, http://www.theglobeandmail.com/commentary/editorials/cellphones-and-cancer-watch-that-thing-by-your-head/article1213907/.
29. Llyod Morgan, "Interphone Studies to Date. An Examination of Poor Study Design Resulting in an Underestimates of the Risk of Brain Tumors," January 23, 2008, http://www.milieuziektes.nl/Rapporten/Interphone%20flaws.pdf.
30. "IARC Director Forces Publication of Interphone Paper," MicrowaveNews.com, May 11, 2009, http://microwavenews.com/Interphone.Wild.html.
31. Camilla, "European Parliament Takes Stand on Health Hazards of Wireless Technologies," ElectromagneticHealth.org, April 21, 2009, http://electromagnetichealth.org/

electromagnetic-health-blog/european-parliament-takes-stand-on-health-hazards-of-wireless-technologies/.

32. Melinda Wenner, "Cellphone Games," *The Walrus*, September 2008, http://thewalrus.ca/cellphone-games/2/?ref=2008.09-health-cellphone-brain-tumour-melinda-wenner&galleryPage.
33. Christopher Wild, "IARC Report to the Union for Internatonal Cancer Control (UICC) on Interphone Study," International Agency for the Research on Cancer and World Health Organization, October 3, 2011, http://interphone.iarc.fr/UICC_Report_Final_03102011.pdf.
34. "IARC Shuts Down Interphone," MicrowaveNews.com, March 18, 2012, http://microwavenews.com/news-center/iarc-shuts-down-interphone.
35. Mukherjee, *The New York Times*, April 13, 2011.
36. M.J. Schoemaker et al., "Mobile Phone Use and Risk of Acoustic Neuroma: Results of the Interphone Case-Control Study in Five North European Countries," *British Journal of Cancer* 93, (2005): 842-848, doi: 10.1038/sj.bjc.6602764.
37. Lennart Hardell and Emily, "Swedish Orebro Studies on Mobile Phone Use Re-Analysed Using the Same Methods as Interphone Study: Increased Risk in Both Studies," ElectromagneticHealth.org, December 17, 2010, http://electromagnetichealth.org/electromagnetic-health-blog/swedish-orebro-re-analysed/.
38. http://www.medicalnewstoday.com/releases/212374.php.
39. Elizabeth Landau, "Study Fails to End Debate on Cancer, Cell Phone Link," CNN.com, May 18, 2010, http://edition.cnn.com/2010/HEALTH/05/14/cell.phones.brain.tumor/index.html?hpt=T2.
40. Siddhartha Mukherjee, "Do Cell Phones Cause Brain Cancer?" *The New York Times*, April 13, 2011, http://www.nytimes.com/2011/04/17/magazine/mag-17cellphones-t.html?pagewanted=all&_r=0.
41. Carl Bialik, "Static in Cellphone Study," *The Wall Street Journal*, May 29, 2010, http://online.wsj.com/article/SB10001424052748704269204575270482215602218.html.
42. Daniel Cressey, "No Link Found Between Mobile Phones and Cancer," *Nature* (2010), doi: 10.1038/news.2010.246.
43. Blake Snow, "Are Cellphone Towers Hazardous to Your Health?" FoxNews.com, August 1, 2011, http://www.foxnews.com/tech/2011/08/01/despite-lack-evidence-alarmist-continue-to-question-cell-phone-safety/.
44. "Report: WHO to Announce Cell Phone, Brain Tumor Link," FoxNews.com, October 26, 2009, http://www.foxnews.com/story/0,2933,569465,00.html.
45. Ibid.
46. Rebecca V. Snowden, "Major Study Complicates Debate Over Cell Phone Use and Cancer Risk," Cancer.org, May 18, 2010, http://www.cancer.org/cancer/news/news/major-study-complicates-debate-over-cell-phone-use-and-cancer-risk.
47. Janet Raloff, "Interphone Study Finds Hints of Brain Cancer Risk in Heavy Cell-Phone Users," ScienceNews.org, May 17, 2010, http://www.sciencenews.org.
48. "Interphone," Public Health England, May 18, 2010, http://www.hpa.org.uk/NewsCentre/NationalPressReleases/2010PressReleases/100518INTERPHONE/.
49. Mona Nilsson, "Conflict of Interest at WHO," May 23, 2011, http://www.monanilsson.se/document/AhlbomConflictsIARCMay23.pdf.
50. http://www.iss.it/pres/prim/cont.php?id=1072&lang=1&tipo=6.
51. http://www.cancer.gov/newscenter/pressreleases/Interphone2010Results.
52. Janet Raloff, "Interphone Study Finds Hints of Brain Cancer Risk in Heavy Cell-Phone Users," ScienceNews.org, May 17, 2010, http://www.sciencenews.org.
53. J. Olsen, "The Interphone Study: Brain Cancer and Beyond," *Bioelectromagnetics* 32, no. 2 (2011): 164-167, doi: 10.1002/bem.20628.
54. Dariusz Leszczynski, "The Newest Epidemiological Study Contains the Same Flaws as the Interphone," *The Washington Times*, January 22, 2012, http://communities.washingtontimes.com/neighborhood/between-rock-and-hard-place/2012/jan/22/newest-epidemiological-study-flaws-INTERPHONE/.
55. http://tobaccodocuments.org/landman/332506.html.

CHAPTER 10: EMF SAFETY STANDARDS

1. J.E. Lovelock, "Midwife to the Greens: The Electron Capture Detector," *Microbiologia* 13, no. 1 (1997): 11–22, http://www.ncbi.nlmf.nih.gov/pubmed/9106178.
2. *Chemical Week*, 16 July 1975, http://www.wunderground.com/resources/climate/ozone_skeptics.asp.
3. Blyskal, J., and M. Blyskal, "PR: How the public relations industry writes the news," New York: William Morrow and Co., 1985, http://www.wunderground.com/resources/climate/ozone_skeptics.asp.
4. Roan, Sharon L., *Ozone Crisis: The 15-year Evolution of a Sudden Global Emergency* (New York: John Wiley & Sons, Inc) 1989, http://www.wunderground.com/resources/climate/ozone_skeptics.asp.
5. Glas, J.P., "Protecting the ozone layer: a perspective from industry," In *Technology and Environment* (ed. by Ausubel, J.H. and H.E. Sladovich), Washington D.C., 1989, http://www.wunderground.com/resources/climate/ozone_skeptics.asp.
6. Cogan, D.G., "Stones in a Glass House," Investor Responsibility Research Center, Washington D.C., 1988, http://www.wunderground.com/resources/climate/ozone_skeptics.asp.
7. Stephen O. Anderson and K. Madhava Sarma, *Protecting the Ozone Layer: The United Nations History* (London: Earthscan, 2002), 297.
8. "Assessment of Health Effects from Exposure to Power-Line Frequency Electric and Magnetic Fields," NIEHS Working Group Report, June 1998, http://www.niehs.nih.gov/health/assets/docs_a_e/emf1.pdf.
9. "Questions and Answers About Electric and Magnetic Fields (EMFs)," U.S. Environmental Protection Agency, December 1992, http://nepis.epa.gov.
10. "ICNIRP Guidelines," *Health Physics* 74, no. 4 (1998): 494–522, http://www.icnirp.de/documents/emfgdl.pdf.
11. "Standards and Guidelines," World Health Organization, http://www.who.int/peh-emf/standards/en/.
12. BIR (2007) 15
13. "Electric and Magnetic Fields (EMF) Radiation from Power Lines," U.S. Environmental Protection Agency, http://www.epa.gov/radtown/power-lines.html.
14. FCC doc: Questions and Answers about Biological Effects and Potential Hazards of Radiofrequency Electromagnetic Fields
15. "Radio Frequency Safety," Federal Communication Commission, http://transition.fcc.gov/oet/rfsafety/rf-faqs.html.
16. FCC doc: Questions and Answers about Biological Effects and Potential Hazards of Radiofrequency Electromagnetic Fields.
17. "FCC's Inadequate Response to Smart Meter Complaint," EMFSafetyNetwork.org, February 4, 2011, http://emfsafetynetwork.org/?p=33851.
18. "Examples of Radiation-Emitting Products," U.S. Food and Drug Administration, http://www.fda.gov/downloads/Radiation-EmittingProducts/ElectronicProductRadiationControlProgram/LawsandRegulations/UCM133545.pdf.
19. "EMFs in the Workplace," Center for Disease Control, NIOSH publication, 1996, http://www.cdc.gov/niosh/docs/96-129/.
20. "Extremely Low Frequency (ELF) Radiation," U.S. Department of Labor, Occupational Safety and Health Administration, https://www.osha.gov/SLTC/elfradiation/index.html.

CHAPTER 11: THE PRECAUTIONARY PRINCIPLE AND THE *BIOINITIATIVE REPORT*

1. "Rio Declaration on Environment and Development," United Nations Environment Program, http://www.unep.org/Documents.multilingual/Default.asp?DocumentID=78&ArticleID=1163.
2. Jonathan H. Adler, "The Problems with Precaution: A Principle without Principle," American.com, May 25, 2011, http://american.com/archive/2011/may/the-problems-with-precaution-a-principle-without-principle.
3. "Framework for Developing Health-Based EMF Standards," Word Health Organization, 2006, http://www.who.int/peh-emf/standards/EMF_standards_framework%5b1%5d.pdf.

4. "European Parliament Recommends Stricter Safety Limits for Cell Phones," MobileTechNews.com, September 19, 2008, http://www.mobiletechnews.com/info/2008/09/19/120132.html.
5. (BIR 2007, Section 1, p. 17).
6. "Comar Technical Information Statement: Expert Reviews on Potential Health Effects of Radiofrequency Electromagnetic Fields and Comments on the Bioinitiative Report," The Committee on Man and Radiation, Health and Physics Society, http://www.emfandhealth.com/12265_COMAR_2009.pdf.
7. Australian Centre for Radiofrequency Bioeffects Research (ACRBR) Position Statement on BioInitiative Report.
8. EPRI comment on The BioInitiative Working Group Report.
9. "MMF Commentary on BioInitiative Report," Australia Mobile Telecommunications Association, http://www.amta.org.au/articles/amta/MMF.commentary.on.BioInitiative.Report.
10. Health Council of the Netherlands Comment on the BioInitiative Report.
11. Ibid.
12. http://www.afsset.fr/upload/bibliotheque/964737982279214719846901993881/Rapport_RF_20_151009_l.pdf.
13. http://www.ncbi.nlm.nih.gov/pmc/articles/PMC2737005/.
14. M Blank (2009) Pathophysiology 16:67–69.
15. Stuart Bramhall, "Electrosmog," January 25, 2011, http://stuartbramhall.aegauthorblogs.com/2011/01/25/electrosmog/.

CHAPTER 12: MINIMIZING EMF RISK

1. "Biological Effects of Power Frequency Electric and Magnetic Fields—Background Paper," U.S. Congress Office of Technology Assessment, May 1989, http://ota.fas.org/reports/8905.pdf.
2. D. Kahneman, *Thinking, Fast and Slow* (2011), 138.
3. Ibid., 141.
4. Betsy Sparrow, Jenny Lui, and Daniel M. Wegner, "Google Effects on Memory: Cognitive Consequences of Having Information at Our Fingertips," *Science* 333, no. 6043 (2011): 776–778, doi: 10.1126/science.1207745.
5. Sherry Turkle, "The Flight from Conversation," *The New York Times*, April 21, 2012, http://www.nytimes.com/2012/04/22/opinion/sunday/the-flight-from-conversation.html?_r=0.
6. "66% of the Population Suffers from Nomophobia," SecureEnvoy.com, February 16, 2012, http://www.securenvoy.com/blog/2012/02/16/66-of-the-population-suffer-from-nomophobia-the-fear-of-being-without-their-phone/.
7. http://www.thefix.com/content/cellphone-addiction-plagues-everyone90088.

CHAPTER 13: CHILDREN AND THE ELECTROHYPERSENSITIVE

1. "IARC Classifies Radiofrequency Electromagnetic Fields as Possibly Carcinogenic to Humans," World Health Organization, May 2011, http://www.iarc.fr/en/media-centre/pr/2011/pdfs/pr208_E.pdf.
2. Gabriel, 2000 (cited in Stewart Report).
3. Christina M. Krause, et al., "Mobile Phone Effects on Children's Event-Related Oscillatory EEG During an Auditory Memory Task," *International Journal of Radiation Biology* 82, no. 6 (2006): 443–450, doi: 10.1080/09553000600840922.
4. Stewart Report, Summary, p8, http://www.iegmp.org.uk/documents/iegmp_1.pdf
5. C.V. Bellieni et al., "Is New Born Melatonin Production Influenced by Magnetic Fields Produced by Incubators?" *Early Human Developement* 88, no. 8 (2012): 707–710, doi: 10.1016/j.earlhumdev.2012.02.015.
6. Joan D. Harland and Robert P. Liburdy, "Environmental Magnetic Fields Inhibit the Antiproliferative Action of Tamoxifen and Melatonin in a Human Breast Cancer Cell Line," *Bioelectromagnetics* 18, no. 8 (1997): 555–562, http://www.ncbi.nlm.nih.gov/pubmed/9383244.

7. Sam Dillon, "Wi-Fi Turns Rowdy Bus into Rolling Study Hall," *The New York Times*, February 11, 2010, http://www.nytimes.com/2010/02/12/education/12bus.html?_r=1.
8. D.K. Li, "A Population-Based Prospective Cohort Study of Personal Exposure to Magnetic Fields during Pregnancy and the Risk of Miscarriage," *Epidemiology* 13, no. 1 (2002): 9–20, http://www.ncbi.nlm.nih.gov/pubmed/11805581.
9. De-Kun Li, Hong Chen, and Roxana Odouli, "Maternal Exposure to Magnetic Fields During Pregnancy in Relation to the Risk of Asthma in Offspring," *Archives of Pediatrics Adolescent Medicine* 165, no. 10 (2011): 945–950, doi: 10.1001/archpediatrics.2011.135.
10. Proceedings International Workshop on EMF Hypersensitivity, http://www.who.int/peh-emf/publications/reports/EHS_Proceedings_June2006.pdf.
11. Magda Havas, "Analysis of Health and Environmental Effects of Proposed San Francisco Earthlink Wi-Fi Network," Trent University, May 31, 2007, http://www.powerwatch.org.uk/pdfs/20061232_havas.pdf.
12. Killer hertz?
13. Olle Johansson, "Aspects of Studies on the Functional Impairment Electrohypersensitivity," *Earth and Environmental Science* 10, no. 1 (2010), doi: 10.1088/1755-1315/10/1/012005.
14. William J. Rea et al., "Electromagnetic Field Sensitivity," *Journal of Bioelectricity* 10, no. 1-2 (1991): 241–256, http://www.aehf.com/articles/em_sensitive.html.
15. Marita Hilliges, Lixin Wang, and Olle Johansson, "Ultrastructural Evidence for Nerve Fibers Within All Vital Layers of the Human Epidermis," *The Journal of Investigative Dermatology* 104, no. 1 (1995): 134-137, doi: 10.1111/1523-1747.ep12613631; Olle Johansson et al., "Intraepidermal Nerves in Human Skin: PGP 9.5 Immunohistochemistry with Special Reference to the Nerve Density in Skin from Different Body Regions," *Journal of the Peripheral Nervous System* 4, no. 1 (1999): 43-52, http://www.unboundmedicine.com/medline/citation/10197064/Intraepidermal_nerves_in_human_skin:_PGP_9_5_immunohistochemistry_with_special_reference_to_the_nerve_density_in_skin_from_different_body_regions_; Lixin Wang et al., "Protein Gene Product 9.5-Immunoreactive Nerve Fibres and Cells in Human Skin," *Cell and Tissue Research* 261, no. 1 (1990): 25-33, doi: 10.1007/BF00329435; Johansson O, Liu P-Y 1995, "Electrosensitivity", "electrosupersensitivity" and "screen dermatitis": preliminary observations from on-going studies in the human skin, In: Simunic, D., ed. Proceedings of the COST 244: Biomedical Effects of Electromagnetic Fields – Workshop on Electromagnetic Hypersensitivity. Brussels/Graz: EU/EC (DG XIII), pp. 52–57.
16. Johansson, *Earth and Environmental Science*.
17. http://www.starweave.com/freiburger/.
18. John Geirland, "The Quiet Zone," *Wired*, February 2004, http://www.wired.com/wired/archive/12.02/quiet.html.
19. Jane O'Brien and Matt Danzico, "'Wi-fi Refugees' shelter in West Virginia Mountains," BBC News, September 12, 2011, http://www.bbc.co.uk/news/world-us-canada-14887428.
20. Ibid.
21. Ibid.

CHAPTER 14: THERAPEUTIC USES OF EMF

1. Alasdair and Jean Philips, "Positive Effects of EMFs," http://www.powerwatch.org.uk/library/downloads/positive-emfs-1-cancer-2012-04.pdf.
2. C. Andrew L. Bassett, "Bioelectromagnetics in the Service of Medicine," *Advances in Chemistry* 250 (1995): 261-275, doi: 10.1021/ba-1995-0250.ch014.
3. Ibid.
4. R.A. Deyo et al., "A Controlled Trial of Transcutaneous Electrical Nerve Stimulation (TENS) and Exercise for Chronic Low Back Pain," *New England Journal of Medicine* 322, no. 23 (1990): 1627–1634, http://www.ncbi.nlm.nih.gov/pubmed/2140432.
5. Per Heden and Arthur A. Pilla, "Effects of Pulsed Electromagnetic Fields on Postoperative Pain: A Double-Blind Randomized Pilot Study in Breast Augmentation Patients," *Aesthetic Plastic Surgery* 32, no. 4 (2008): 660–666, http://link.springer.com/article/10.1007/s00266-008-9169-z.
6. http://www.sciencedirect.com/science/article/pii/S1067251697801184; Brock Liden,

Melitta Simmons, and Adam S. Landsman, "A Retrospective Analysis of 22 Patients Treated with Percutaneous Radiofrequency Nerve Ablation for Prolonged Moderate to Severe Heel Pain Associated with Plantar Fasciitis," *Journal of Foot and Ankle Surgery* 48, no. 6 (2009): 642–647.

7. I. George et al., "Myocardial Function Improved by Electromagnetic Field Induction of Stress Protein Hsp70," *Journal of Cellular Physiology* 216, no. 3 (2008): 816–823, http://www.unboundmedicine.com/medline/citation/18446816/Myocardial_function_improved_by_electromagnetic_field_induction_of_stress_protein_hsp70_.
8. Ibid.; H. Lin, M. Blank, and R. Goodman, "Magnetic Field-Responsive Domain in the Human HSP70 Promoter," *Journal of Cellular Biochemistry* 75, no. 1 (1999): 170–176, http://www.ncbi.nlm.nih.gov/pubmed/10462715.
9. K.T. Hartner and D. Pette, "Fast and Slow Isoforms of Troponin I and Troponin C. Distribution in Normal Rabbit Muscles and Effects of Chronic Stimulation," *European Journal of Biochemistry* 188, no. 2 (1990): 261–267, http://www.ncbi.nlm.nih.gov/pubmed/2318209.
10. "Would You Believe . . . Specific Frequencies Block Growth of Cancer Cells," MicrowaveNews.com, December 1, 2011, http://microwavenews.com/ModulationFrequencies.html.
11. http://www.ncbi.nlm.nih.gov/pubmed/7744555; Int J Neurosci, November 1994 79(1–2):99–110. A drug naive parkinsonian patient successfully treated with weak electromagnetic fields.
12. R.E. Hoffman et al., "Transcranial Magnetic Stimulation and Auditory Hallucinations in Schizophrenia," *The Lancet* 355, no. 9209 (2000): 1073–1075, http://www.ncbi.nlm.nih.gov/pubmed/10744097; J. Horacek et al., "Effect of Low-Frequency rTMS on Electromagnetic Tomography (LORETA) and Regional Brain Metabolism (PET) in Schizophrenia Patients with Auditory Hallucinations," *Neuropsychobiology* 55, no. 3–4 (2007): 132–142, doi: 10.1159/000106055, http://www.karger.com/Article/Abstract/106055
13. Nuray Yozbatiran et al., "Safety and Behavioral Effects of High-Freqency Transcranial Magnetic Stimulation in Stroke," *Stroke* 40 (2009): 309–312, doi: 10.1161/STROKEAHA.108.522144.
14. Peter Eichhammer et al., "Neuronavigated Repetitive Transcranial Magnetic Stimulation in Patients with Tinnitus: A Short Case Series," *Biological Psychiatry* 54, no. 8 (2003): 862–864, http://www.sciencedirect.com/science/article/pii/S0006322302018966.
15. J.E. Kenkre et al., "A Randomized Controlled Trial of Electromagnetic Therapy in the Primary Care Management of Venous Leg Ulceration," *Family Practice* 13, no. 3 (1996): 236–241, doi: 10.1093/fampra/13/3/236.

CHAPTER 15: THE NEXT STEP

1. http://www.unep.org/Documents.Multilingual/Default.asp?documentid=78&articleid=1163
2. A. Kong et al., "Rate of De Novo Mutations and the Importance of Father's Age to Disease Risk," *Nature* 488, no. 7412 (2012): 471–475, doi: 10.1038/nature11396.

ABOUT THE AUTHOR

Martin Blank graduated from the Bronx High School of Science and City College of New York, and received PhDs from Columbia University and Cambridge University. He is currently a Special Lecturer in the department of physiology and cellular biophysics at Columbia University, where he was Associate Professor for over forty years. During this period, he edited twelve books and published over two hundred papers on the biological effects of EMF. He has worked for the Office of Naval Research, as well as several industrial labs, including Unilever, and is regularly asked to be an expert witness on cases involving the effects of EMF. He has also organized scientific meetings, including two World Congresses on Electricity and Magnetism in Biology and Medicine, and the Gordon Research Conference on Bioelectrochemistry. Additionally, he has served on the editorial boards of several journals, including *Electromagnetic Biology and Medicine*, and edited the 2009 issue of *Pathophysiology* on biological effects of EMF. He was one of the organizers of the *Bioinitiative Report*, where he authored the section on stress proteins. He has also served as the President of the Bioelectromagnetics Society.

ABOUT SEVEN STORIES PRESS

Seven Stories Press is an independent book publisher based in New York City. We publish works of the imagination by such writers as Nelson Algren, Russell Banks, Octavia E. Butler, Ani DiFranco, Assia Djebar, Ariel Dorfman, Coco Fusco, Barry Gifford, Martha Long, Luis Negrón, Hwang Sok-yong, Lee Stringer, and Kurt Vonnegut, to name a few, together with political titles by voices of conscience, including Subhankar Banerjee, the Boston Women's Health Collective, Noam Chomsky, Angela Y. Davis, Human Rights Watch, Derrick Jensen, Ralph Nader, Loretta Napoleoni, Gary Null, Greg Palast, Project Censored, Barbara Seaman, Alice Walker, Gary Webb, and Howard Zinn, among many others. Seven Stories Press believes publishers have a special responsibility to defend free speech and human rights, and to celebrate the gifts of the human imagination, wherever we can. In 2012 we launched Triangle Square books for young readers with strong social justice and narrative components, telling personal stories of courage and commitment. For additional information, visit www.sevenstories.com.